PATENT POLITICS

Patent Politics

Life Forms, Markets, and the Public Interest in the United States and Europe

SHOBITA PARTHASARATHY

THE

UNIVERSITY

OF CHICAGO

PRESS

Chicago and London

The University of Chicago Press, Chicago 60637

The University of Chicago Press, Ltd., London

© 2017 by The University of Chicago

Published 2017

Paperback edition 2020

Printed in the United States of America

29 28 27 26 25 24 23 22 21 20   1 2 3 4 5

ISBN-13: 978-0-226-43785-9 (cloth)

ISBN-13: 978-0-226-75913-5 (paper)

ISBN-13: 978-0-226-43799-6 (e-book)

DOI: https://doi.org/10.7208/chicago/9780226437996.001.0001

Library of Congress Cataloging-in-Publication Data

Names: Parthasarathy, Shobita, author.

Title: Patent politics : life forms, markets, and the public interest in the United States and Europe / Shobita Parthasarathy.

Description: Chicago : The University of Chicago Press, 2017. | Includes bibliographical references and index.

Identifiers: LCCN 2016026515 | ISBN 9780226437859 (cloth : alk. paper) | ISBN 9780226437996 (e-book)

Subjects: LCSH: Biotechnology—Patents. | Patent laws and legislation—United States—History. | Patent laws and legislation—Europe—History. | Bioethics—United States. | Bioethics—Europe.

Classification: LCC TP248.175 .P35 2017 | DDC 174.2–dc23

LC record available at https://lccn.loc.gov/2016026515

♾ This paper meets the requirements of ANSI/NISO Z39.48–1992 (Permanence of Paper).

CONTENTS

ACRONYMS AND ABBREVIATIONS

AAVS	American Anti-Vivisection Society
ACLU	American Civil Liberties Union
ALDF	Animal Legal Defense Fund (US)
AMP	Association for Molecular Pathology (US)
APA	Administrative Procedures Act (US)
ASPCA	American Society for the Prevention of Cruelty to Animals
BPD	Biotech Patent Directive: formally, the EU Directive for the Legal Protection of Biotechnological Inventions
BRCA1/2	Breast Cancer Genes 1 and 2
CAFC	Court of Appeals for the Federal Circuit (US)
CCPA	Court of Customs and Patent Appeals (US)
CCST	California Council on Science and Technology (US)
cDNA	complementary DNA
CGS	Center for Genetics and Society (US)
CIRM	California Institute for Regenerative Medicine (US)
Commission	European Commission
Council	Council of Europe
DOJ	Department of Justice (US)
EBA	Enlarged Board of Appeal (Europe)
ECJ	European Court of Justice
ECLJ	European Center for Law and Justice
EEC	European Economic Community
EFTA	European Free Trade Association
EGE	European Group on Ethics in New Science and Technologies
EPA	Environmental Protection Agency (US)
EPC	European Patent Convention
epi	Institute of Professional Representatives before the European Patent Office (European Patent Institute)
EPO	European Patent Office
ESC	Economic and Social Committee (Europe)
EST	Expressed sequence tag
EU	European Union
FDA	Food and Drug Administration (US)

FET	Foundation on Economic Trends (originally the People's Bicentennial Commission, then the People's Business Commission, and now FET) (US)
FTCR	Foundation for Taxpayer and Consumer Rights (now Consumer Watchdog) (US)
GAEIB	Group of Advisors on the Ethical Implications of Biotechnology (Europe)
hESC	Human embryonic stem cells
HGP	Human Genome Project
HHS	Department of Health and Human Services (US)
ICTA	International Center for Technology Assessment (US)
KPAL	Kein Patent Auf Leben/No Patents on Life (Europe)
MEP	Member of the European Parliament
Myriad	Myriad Genetics (US)
NIH	National Institutes of Health (US)
NPOS	No Patents on Seeds (Europe)
NRC	National Research Council (US)
Organization	European Patent Organization
OSGATA	Organic Seed Growers and Trade Association (US)
Parliament	European Parliament
PBC	People's Business Commission (originally the People's Bicentennial Commission, now FET) (US)
PTO	Patent and Trademark Office (US)
PubPat	Public Patent Foundation (US)
RAC	Recombinant DNA Advisory Committee (US)
rDNA	recombinant DNA
SACGHS	HHS Secretary's Advisory Committee on Genetics, Health, and Society (US)
SACGT	HHS Secretary's Advisory Committee on Genetic Testing (US)
SAWS	Sensitive Application Warning System (US PTO)
SCOTUS	US Supreme Court
SeCa	Sensitive Cases system (EPO)
SftP	Science for the People (US)
TBA	Technical Board of Appeal (Europe)
TRIPS	Trade-Related Intellectual Property Rights Agreement
UPOV	International Union for the Protection of New Varieties of Plants
USDA	US Department of Agriculture
WARF	Wisconsin Alumni Research Foundation (US)

INTRODUCTION

The headquarters of the European Patent Office (EPO) is not a place known for political activism. And yet, on a cool morning in October 2013, dozens of protestors barricaded themselves in front of the EPO's main buildings in Munich and demanded that the patent bureaucracy "Stop Patents on Life." These civil society challengers—including farmers, environmentalists, representatives of international development organizations, and average citizens—had gathered to oppose the EPO's decisions to permit patents on conventional plants and animals bred using biotechnology, which, they argued, would accelerate dangerous consolidation in the agricultural industry.[1] Equally important, they suggested, such patents would continue to disempower small farmers, who would be forced to depend on patent holders for their crops and ultimately have trouble maintaining independent businesses. But underlying all of these arguments lay concerns that the EPO was, as their signs put it, a "democracy-free zone". Protestors charged that the patent bureaucracy did not adequately consider the will of the European people, but instead privileged private interests while cloaking its decisions in technical and legal language.

Similarly, a few months earlier and an ocean away, breast cancer advocates had stood on the steps of the United States Supreme Court decrying the US Patent and Trademark Office's (PTO) policy of allowing patents on human genes.[2] They, along with scientists, physicians, genetic counselors, and civil rights advocates, argued that gene patents were immoral and illegal because they commodified nature. Moreover, they noted, such patents made it difficult for biologists to conduct fundamental research and for patients to access the best health care. Instead of improving the public's access to innovation, these plaintiffs charged, human gene patents did the opposite. They would prevail in June 2013, when the Supreme Court ruled in *AMP v. Myriad* that isolated human genes were unpatentable.

In recent decades, we have seen hundreds of civil society–led challenges to the world's science and technology policymaking institutions, signaling growing citizen distrust. As science and technology have become ubiquitous in our daily lives, they are having noticeably positive effects. But citizens have also begun to wonder whether the systems that regulate the development, availability, and use of innovation reflect their values and concerns. Some, like the environmentalists and patient advocates mentioned above, are even mobilizing in opposition and seeking a greater voice in policymaking. They are no longer content to accept the judgment of technical experts regarding the implications, including the risks and ethics, of innovation. At first glance, patent systems seem an unlikely place for these controversies. Historically they have been technical and esoteric domains operating far from public view, of real interest only to those who seek to gain an exclusive right to commercialize their inventions. Increasingly, however, they are fraught sites, linked to issues of health, economic inequality, and morality. How are governments handling these challenges? What do they mean for the future of patent systems, and innovation systems more broadly? What do these controversies tell us about the contemporary politics of science and technology? And how might they help us develop better strategies for governing innovation?

To explore these questions, I focus on the most sustained set of these challenges: those concerning patents on life-forms, including genetically modified organisms, human genes, and stem cells, in the United States and Europe. These patents have been extremely controversial in both places for more than thirty years. But in the United States and Europe the battles have proceeded differently and produced different responses, despite many political and economic similarities between the two places, which include joint efforts to facilitate a global marketplace through international intellectual property treaties. The United States has opened its doors wide to life-form patentability, famously allowing patents on "anything under the sun made by man" in the 1980 *Diamond v. Chakrabarty* case.[3] And it has largely dismissed ethical, socioeconomic, health, and environmental concerns, characterizing them as distractions in a domain focused on technical questions of novelty, utility, and inventiveness. The only relevant question, US decision makers suggest, is whether an innovation is a novel technology or simply a natural discovery. Europe, by contrast, has taken these concerns seriously: it now excludes multiple biotechnologies from patentability on ethical and socioeconomic grounds. Scholars and other observers tend to explain these differences in terms of the law: the United States has developed a "product of nature" doctrine in its case precedent that excludes natural phenomena from patentability, for example, while European laws prohibit patents on inventions whose publication or exploitation might be contrary to public order or morality (known as *ordre public*).[4]

This legal explanation, however, is insufficient. Why did European law and patent evaluation evolve in a manner that allowed those concerned with the ethics of biotechnology to use the *ordre public* clause to shape patent policy? Why, in the case of the United States, have similar ethical concerns failed to give rise to comparable changes in the patent system? After all, US courts have recognized a "moral utility" doctrine in patent law.[5] Why did it not figure in the US discussions regarding biotechnology patents? Why did even the challenges themselves look different in the United States and Europe? And what were the implications for the way the two jurisdictions thought about the governance of both patents and innovation generally?

Throughout this book, I explain these differences in how the United States and Europe navigated concerns about the implications of life-form patents by demonstrating that patents are social and political, not just technical and legal, achievements. The United States and Europe have long thought quite differently about the relative roles of the government and the market in shaping innovation and achieving the public interest. This, I demonstrate, has guided the development of both the political environments and legal frameworks of their patent systems, which in turn has shaped the way they understand patents. To make this argument, I compare the politics of the US and European patent systems from their origins through the recent tumult brought about by biotechnology. This approach reveals the often-overlooked differences in the structures, orientations, and ongoing political work of supposedly similar modern patent systems, which lie underneath different definitions of patents in the United States and Europe. It also allows us to see the usually hidden implications of these different understandings, which go far beyond the world of biotechnology patent law. As the United States and Europe debated the patentability of microorganisms, genetically modified plants and animals, human embryonic stem cells, and human genes, they developed rather different approaches to patent and innovation governance. This included different ideas about the roles and responsibilities of patent system institutions, different understandings of who should participate and how they should do so, and different perspectives on how life itself should be governed.

BEYOND TECHNOLOGY AND THE LAW

We tend to think of patents as global symbols of invention.[6] They are issued through what seems to be an objective and straightforward process: technically trained bureaucrats across the world award them to developers of new and useful technologies after examining related scientific publications and patents (known as "prior art"). The patents themselves are legal certifications that provide their owners with the right to prevent others from commercializing the invention for a limited period of time. Others who wish to commercialize the invention must purchase a license from the patent holder. And patents seem to be generally recognized across jurisdictions. Indeed, the legal rules that guide patent systems, including the criteria for patentability, the scope of the rights patents confer, and the structures and processes of patent systems have been harmonized through international agreements that date back more than a century.

But in recent years scholars have reminded us that a patent's meaning is shaped by its context. It has changed over time, with early rewards for individual entrepreneurship becoming, by the twentieth century, lucrative tools that have shaped the business strategies of high-tech industries.[7] As the number of patents has increased, and the scope of patentability has grown, they have slowly become tools to increase the revenues of universities as well. Some argue that this has transformed academic culture and, specifically, the practice of science.[8] Today, patents are increasingly viewed as double-edged: they are not just important innovation incentives; they can also be barriers to future research.[9]

Patents also function differently across the globe. In the United States and Europe, traditional participants in the patent system—including inventors, patent lawyers, and patent bureaucracies—see patents primarily as legal and economic weapons to prevent others from commercializing a particular invention and to sue others for infringement. But in Africa, for example, where market players are unlikely to have the funds to apply for and defend a patent, or purchase the exclusive license to commercialize a patented innovation, they are often treated as sources of technical information rather than bundles of legal rights.[10] African intellectual property offices stabilize this definition by publishing monographs that describe inventions that might inspire local innovation among local farmers and industries.

Patents do more than influence innovation trajectories and business strategies. As they explicitly define patentable "invention," they are also shaping social understandings of nonpatentable "discovery" and "nature."[11] In helping to generate markets for new technologies, they emphasize particular kinds of economic relationships at the expense of others.[12] Finally, researchers have found that the descriptions contained within the patents themselves can shape how we define social categories such as race and citizenship.[13]

Despite this growing attention to the role and meaning of patents, however, we know little about their political machinery. Legal scholars have taught us about the national, regional, and international legal frameworks that structure patent systems, but how are these frameworks implemented and by whom? What kinds of assumptions and logics structure the workings of patent systems, and what are the implications for the governance of innovation? Who shapes patent system decision making and policy, and what claims to knowledge and expertise do they use to justify their participation? How does political culture and ideology figure in these structures and processes? What understanding of the public interest do patents embody?

We might assume that the answers to these questions are clear, particularly given the technical and legal orientations of patent systems. But they are not. By comparing the controversies over US and European life-form patents, which have been particularly long and comprehensive, we can bring these politics into sharp relief. As a wide variety of citizens — from scientists to animal-rights activists — have asserted their right to participate in this highly specialized expert domain and to introduce concerns that usually go unrecognized, they have revealed the values and assumptions that lie behind seemingly objective patent systems and that structure their rules, practices, institutional configurations, and decision making. We will see how political culture and ideology have shaped this machinery and explore the consequences for patent and innovation governance. This analysis should cause us to rethink the supposedly similar understandings of patents across jurisdictions, reminding us that even highly technical areas of law and policy are shaped in fundamental and important ways by their contexts. As a result, it should force us to consider values and politics much more seriously in our ongoing patent reform efforts. And, it should lead us to question the meaning and suitability of international legal agreements in providing the scaffolding for an interconnected world.

CHALLENGING INNOVATION SYSTEMS

The life-form patent controversies are part of a long legacy of organized, grass-roots critiques of innovation systems that have emerged across the world since the 1960s. In the 1940s and 1950s, governments began to increase their investments in science and technology, which seemed to have demonstrated social benefits in the form of national security, life-saving medical treatments, more efficient agriculture, and transnational communication, just to name a few.[14] But as civil rights and other social movements blossomed, citizens began to question whether these science and technology investments were really in the public interest, including whether they were morally appropriate, addressed or possibly exacerbated economic inequalities, and protected the environment. Scientists led some of this activism, calling attention to their own complicity in military research and technology as well as in environmental degradation.[15] In her book *Silent Spring*, Rachel Carson demonstrated how DDT, a pesticide used to combat malaria, typhus, and other insect-borne human diseases, had toxic effects on wildlife and the ecosystem.[16] Carson's findings stimulated

5

more pesticide regulations and eventually the withdrawal of DDT. But they also galvanized a global movement that urged greater government attention to the environmental implications of new technologies.[17] Soon after, worried about the costs of nuclear power plants, the health and environmental impact of nuclear waste disposal and accidents, and the possibility of terrorist attacks, both American and European citizens mobilized against nuclear power.[18] By the 1990s, activists from Paris to Peru had begun to protest the rise of biotechnology, particularly with regard to the health, environmental, and socioeconomic risks of genetically modified crops.[19] Meanwhile, patients had begun to question whether government decisions regarding both biomedical research funding and the regulation of medical interventions really reflected their priorities.[20] The idea that the priorities of scientists and patients might conflict was a revelation. Overall, these challenges to research and regulatory policy-making across the world have forced governments to modify their decision-making processes regarding science and technology, including creating new opportunities for public engagement.[21]

Challenges to the patent system began in the 1970s. Activists focused initially on the new "life forms"—from DNA fragments to genetically engineered organisms—that recombinant DNA (rDNA) techniques had begun to produce and that would surely increase with the rise of biotechnology. They worried about the transformation of life not only into an object of research, ripe for manipulation, but also into a commodity, to be bought and sold like any other patented technology in the marketplace. These concerns emerged first in the United States, in the *Diamond v. Chakrabarty* case. Ananda Chakrabarty, a scientist at General Electric, applied for a patent covering a microorganism genetically engineered to eat oil. Unsure whether this microorganism was unpatentable "nature" or patentable "technology," the US Patent and Trademark Office (PTO), led by Commissioner Sidney Diamond, rejected the application.[22] General Electric appealed, and the case eventually landed at the Supreme Court. As in many other Supreme Court patent cases, patent lawyers, scientific organizations, individual scientists, and universities weighed in, in the form of amicus (friend of the court) briefs. All of them supported patentability, citing biotechnology's enormous economic, medical, and agricultural promise. Jeremy Rifkin, known today as a biotechnology watchdog, and his allies filed the lone opposing brief.[23] Any third party can submit an amicus brief to the Supreme Court, but this was the first time that civil society representatives had filed one in a Supreme Court patent case. It was challenging to do: in order to capture the attention of the justices and their clerks, briefs usually used legal language and referred to patent-law doctrine, which was unfamiliar to activists.[24] In the end, the Rifkin coalition argued that patents on life forms were unethical because they would introduce a new form of ownership and transform the relationship between humans and their environment, and because the monopolies created by such patents would lead eventually to agricultural consolidation and reduced biodiversity. With these arguments came implicit and explicit calls to treat life-form patentability as a moral, rather than a technical, question that required discussion among policymakers and the public, and the consideration of a wider array of knowledge and expertise than what was customary in the patent system.

In the decades since, citizens have raised concerns similar to Rifkin's across a wide swath of technologies, from pharmaceuticals to the large-scale technologies designed to mitigate climate change known as geoengineering.[25] They question whether patents in these areas are indeed in the public interest, noting that the control over technologies that patents give to their owners can create market incentives where there should be none, drive development in morally and socially deleterious ways, and limit access to socially important innovation. Governments must, they argue, step in to consider these issues and protect the public interest. These worries have traveled across borders, affecting both domestic politics and negotiations over international trade agreements.[26] Meanwhile, the controversies over life-form patents have endured, likely due to the ongoing excitement and trepidation associated with the field of biotechnology. While it has stimulated thousands of patent applications across the world, biotechnology continues to raise fundamental questions about the meanings of nature and of life, and about the appropriate limits of tinkering with our bodies, our ecosystems, and our food supply.[27]

As the life-form patent controversies have evolved, they have reflected this broadening set of concerns. First, challengers argue that patents produce monopolies that exacerbate inequalities rather than increase access to innovation. This includes reduced access to important medical diagnostics and treatments.[28] They also observe that patents can damage health care quality; as they launched their case at the US Supreme Court, patient advocates, scientists, health care professionals, and civil liberties groups noted that human, and particularly breast cancer (BRCA), gene patents had created monopolies that eliminated the possibility of confirmatory diagnostic testing.[29]

Second, many question how patent systems assign and recognize value in innovation. The rules and processes associated with applying for a patent, they argue, reward scientific and technical work but do not recognize the value of other contributors—from taxpayers to research subjects—who facilitate the research.[30] For example, governments and civil society groups recognize knowledge from indigenous communities, especially those in the developing world, for informing drug and agricultural development in significant ways.[31] But these communities rarely receive official credit, because their knowledge is published neither in scientific journals nor in patents but is passed down either orally or in nonscientific texts.

A third set of concerns focuses on the possibility that patents actually stifle innovation. Patents at early stages of research and development become unpredictable and time-consuming hurdles to jump over. In 2014, billionaire entrepreneur Elon Musk famously gave up dozens of patents related to electric vehicles in order to stimulate innovation in the field.[32] In the case of biotechnology, early-stage patents have been viewed as particularly problematic, especially when the patented life form is itself the object of research, as in the cases of human embryonic stem cells and human genes.[33] Critics argue that such hurdles can have effects long after the twenty-year life of a patent is over, because it privileges certain research teams and trajectories of innovation.

Finally, some worry that patents encourage innovation and markets in morally questionable areas. In the 1980s and 1990s, farmers and animal-rights activists worried that patents on genetically modified animals would encourage research that would lead to unnecessary animal suffering.[34] More recently, civil society groups have questioned the wisdom of patents covering the creation of novel organisms produced using synthetic biology techniques.[35] Patents, they argue, could encourage private investment in the development of these new organisms without adequate public debate over whether such organisms should be produced in the first place. Such organisms, they suggest further, could transform our understanding of life, force humans to engage in unethical behaviors, and, if released, harm our ecosystems. Taken together, these criticisms assume that patent systems bear responsibility far beyond the legal certification of inventions. They envision, in other words, the system's scope to include issues of public morality, public health, economic inequality, environmental sustainability, and civil rights.

As I noted above, the United States and Europe have adjudicated these concerns differently. But what is particularly puzzling, given their many similarities, is that they have interpreted the issues at stake themselves, and their meaning in the patent system, quite differently too. In the United States, legislators, judges, patent lawyers, and bureaucrats tend to argue that most of these concerns are misguided. Moral, social, distributive, health, or environmental issues are seen to be the province of health or agricultural regulators, not patent policymaking. The system is very limited, these decision makers state, and only confers a "negative right" that allows inventors to prevent others from commercializing their inventions. The only relevant concern, they suggest, is whether patents stifle or stimulate innovation. And on that point, they argue, the evidence is unclear.[36] Meanwhile, European patent-system institutions tend to view patents as having wide-ranging implications. The certification of inventions that patents represent, they argue, is a moral act, and must be treated as such.[37] Furthermore, they see the patent system as responsible for the availability and use of patented technology. And in the wake of the life-form patent controversies, they have begun to transform their rules, processes, and institutions with this understanding guiding them.

What is treated as technical and legal in the US patent system, then, is seen as moral and social in the European patent system. Why? And how do these differences take institutional shape and persist, particularly in the face of international legal harmonization? Throughout this book, I argue that these differences are the result of different patent system logics, which are rooted in different political cultures and ideologies vis-à-vis innovation policy. By *political culture*, I mean the "systematic means by which a political community makes binding collective choices."[38] And by *ideology*, I am referring to collectively held beliefs about how government should function in society that form a coherent, understandable picture.[39] But political culture and ideology alone do not set patent systems on a clear or easily predictable path. Along the way, institutions and organized groups do enormous work to bring these logics into policymaking and practice, and to help them take organizational form, often in unforeseeable ways. My goal in this book is to reveal how these logics shape a patent system's structure, its practices, and its political environment. Ultimately, they influence patent-system decisions and even our understandings of patents.

Unlike many scholars of patents, then, I view them in social, political, and historical terms. Therefore, I first ground my analysis in the histories of the US and European patent systems. Drawing from laws, court cases, legislative hearings, and other primary and secondary sources, I reconstruct how each system interpreted the patent system's role and commitment to the public interest. And I demonstrate how these interpretations became logics that drove the development of patent system rules, practices, institutional configurations, and the rhetorics of decision makers and traditional organized interests.

Then, as I turn to the life-form patent controversies, I explore them in political as well as legal terms. This means that as I explore how each jurisdiction contends with questions of patentability, I delve deeply into the assumptions, rhetoric, formal rules, informal practices, institutional structures and politics, and organized interests that constitute each patent system's political order and shape its law and understanding of patents. To do this, I rely on more than legal decisions and official hearing transcripts. I conducted more than one hundred interviews with US and European government officials (primarily those at patent offices), patent lawyers, patent holders, and civil society groups and citizens initiating challenges to the life-form patents. I also attended legal hearings, workshops, and informal meetings related to life-form patents and patent governance in the two places, which positioned me to understand both the legal issues at stake and the institutional politics animating these domains. In addition, I analyzed documents related to the controversies produced by patent offices, the courts, legislatures, patent holders, and civil society challengers. This included the "prosecution," all of the documents filed at the patent office in relation to a particular patent, which usually ran into the thousands of pages. Finally, when I do include legal texts, I interpret them somewhat differently than a legal scholar would. While the legal scaffolding is important to me, I focus on how the texts embody a particular worldview and set of priorities.

Overall, this data opens a window into the size and scope of the life-form patent controversies in both places, and helps us understand how and why these wide-ranging battles were disciplined as they churned through each patent system. As such, it offers a fresh lens for interpreting legal outcomes as well as a more comprehensive understanding of the reach of political culture and ideology in shaping science and technology and related policy domains.

COMPARING THE US AND EUROPEAN PATENT SYSTEMS

The United States and Europe are particularly good sites for this comparative analysis because of their many political and economic similarities: they are both democratic and capitalist in orientation, and both have long valued intellectual property rights as a means of achieving economic growth. In addition, after one hundred years of international legal harmonization, their patent systems generally look alike.

Efforts to create a global intellectual property regime began in 1883, with the creation of the Paris Convention, which established some basic rules, including treating patent applicants from all participant countries the same.[40] By 1900, the convention had fifteen signatories, including the United States and ten countries across Europe. Then, in the wake of the world wars, the United Nations developed the World Intellectual Property Organization (WIPO) to provide a "global forum for intellectual property services, policy, information, and cooperation."[41] In 1970, eighteen countries, including the United States and much of Europe, created a unified patent-filing system by signing WIPO's Patent Cooperation Treaty. When most Western European countries created a pan-European patent system in 1973, in order to strengthen the European economy as well as European peace, they modeled it on their US counterpart. Called the European Patent Organization, this pan-European patent system—which is a focus of this book—now works with its counterparts in the United States and Japan to ensure "an increasingly efficient worldwide patent system in the 21st century" through the Trilateral Co-operation.[42] The Trilateral organizes personnel exchanges, joint patent-review exercises, and frequent meetings and discussions to facilitate harmonization. In other words, it tries to reinforce the legal similarities through harmonization of its technical practices. And, as recently as 2011, the United States Congress passed a law to ensure that its patent system looked like its European counterpart's, in terms of both its filing system and its opportunities for dispute resolution after a patent is granted.[43]

As a result of this long-standing cooperation, the US and pan-European patent systems appear similar in both legal and structural terms. In fact, scholars often refer to a "Euro-American" approach to intellectual property that contrasts with those in the developing world.[44] In both jurisdictions, patent bureaucracies, populated with technically and legally trained personnel, play a central role by examining patent applications and granting patents.[45] Examiners usually have an educational background in engineering or the natural or physical sciences, and receive some on-the-job training in patent law. After a period of back-and-forth with an inventor, the examiner decides whether or not to issue a patent on the basis of criteria that are substantially similar in the United States and Europe.[46] In both places, the patent is valid for twenty years. Both systems have additional legal oversight, allowing the inventor to dispute the patent office's decision, another innovator to dispute the invention's patentability, and the patent holder to sue infringers. Legislative bodies supervise both systems, enacting and reviewing the laws that guide the patent bureaucracies. The organized interests that populate both jurisdictions have been largely similar in the United States and Europe, including inventors and their employers as well as patent lawyers and the technically trained agents who assist them. Finally, the United States and Europe see similar numbers of patent applications. In 2014, out of the more than 2.6 million applications across the world, the US PTO received more than five hundred thousand applications, while national patent offices across Europe received approximately three hundred and fifty thousand, and the EPO alone received over one hundred and fifty thousand.[47]

There are, however, a few institutional differences between the two patent systems. In the United States, Congress and the courts oversee the Patent and Trademark Office, which is nestled inside the executive branch of government. In fact, the PTO is particularly weak in comparison with regulatory agencies like the Food and Drug Administration and the Environmental Protection Agency (EPA): it cannot change its own governing statutes—only Congress can.[48] By contrast, the European Patent Organization is a stand-alone, pan-governmental organization, with thirty-seven member countries. It is not officially connected to the European Union. While it often chooses to abide by the laws of the European Union in order to facilitate cross-border trade, it is under no official obligation to do so.[49] The Organization includes the EPO, which reviews and grants patent applications, and an Administrative Council that serves as its legislative body and includes representatives from all of its member countries.

The European Patent Organization works in parallel with national European patent systems, which function within national governments. The European Patent Office examines patent applications and grants patents, but it did not issue a common European patent during the period analyzed in this book. (The Organization's member states have now made a preliminary agreement to create a unitary patent across Europe.)[50] Rather, if the EPO decides to grant a patent after its examination process, it grants national patents valid in its member countries. (Applicants decide in advance where they would like patents granted, and pay fees on a sliding scale according to the number of countries.) Throughout its history the Organization has grown considerably and is now the primary recipient of patent applications in Europe.[51] Since 1980, EPO applications have grown more than 800 percent, while applications to the UK Patent Office, for example, have diminished by almost half.[52] Today, innovators who seek a patent only in one country usually apply in a national patent office. But those who want a patent valid in multiple European countries likely submit their applications through this pan-European system. And, as we will see, the European Patent Organization's increasing importance has made it the central site of the European life-form patent controversies.

Although these differences suggest that the European patent system could be more politically exposed than its US counterpart, I argue throughout the book that, at a deeper level, the differences between the US and European approaches to patents and their governance are the result of political culture and ideology. The United States tends toward an adversarial style of politics, with policy often the result of a battle among interest groups in both policy-making and public spaces.[53] European countries have a reputation as more "corporatist": decisions are often made through consensus-building among disparate—but a more limited number of—groups.[54] In addition, European countries, with a long history of government involvement in, and often spon-sorship of, health care systems and pension plans, have a different approach to social welfare than the United States.[55] Finally, scholars have identified dif-ferent "civic epistemologies"—institutionalized practices by which citizens "test knowledge claims used as a basis for making collective choices"—in the United States and European countries.[56]

But particularly important in shaping their patent systems, I argue, are US and European political ideologies regarding the role of government vis-à-vis the market, innovation, and morality. While many assume that the United States and Europe have similar commitments to neoliberalism, scholars have demonstrated that the two places do not always interpret their capitalist com-mitments in the same way.[57] I suggest that the United States has a *market-making* approach to innovation, which values an unfettered market in order to achieve overall social benefit. Furthermore, the free market, through the rational, collective decisions of both producers and consumers, is generally thought to address moral concerns. In other words, the amoral marketplace produces moral order: if an item is made available in the marketplace and pur-chased in sufficient quantities that it continues to be available, then the item is morally unproblematic and socially beneficial. A powerful marketplace is also believed to encourage an innovative and entrepreneurial spirit, as well as, eventually, more beneficial technology. The role of government, in this view, is to facilitate market activities and to ensure that technologies are produced and brought into commerce to be valued and purchased.

Of course, the US government does intervene in innovation policymaking in multiple ways.[58] But these interventions are usually justified in a way that underscores a market-making ideology: governments should intervene on the occasions when the activities of the market alone fail in producing the public interest. This includes the US patent system itself, which could be interpreted as artificial interference in the free flow of market activities. But, the potential benefits of the patent system, which include the stimulation of innovation, the reduction of trade secrets, the availability of more technology, and ulti-mately the expansion of the marketplace, are seen as outweighing the dangers of time-limited monopolies. Indeed, even the idea that government should intervene only in the case of "market failure" assumes that the market can generally be trusted to produce decisions in the public interest. This explains why bioethics committees in the United States are invariably valued for shaping public debate related to biomedicine and biotechnology—and thereby producing more rational actors—but are rarely influential in producing regulatory policy: the consumer, not the bioethicist, is the ultimate moral actor.[59]

Europe, by contrast, takes what I call a *market-shaping* approach. While a pan-European political culture is only now emerging (and, as I document through the book, the patent system controversies are helping to shape it), most Western European countries have traditionally valued a more active role for government in regulating innovation and shaping its moral, social, and economic implications. Moral order is seen as prior to and larger than the marketplace, and both markets and innovation are seen as potentially morally ambiguous. Markets are not the natural state, but rather an activity in which the moral community participates. And in this activity, the interests of innovators and the public are not necessarily the same. Put differently, citizens do not trust markets to produce moral and social benefits automatically. They expect the state to create and maintain this moral order, and to minimize the harms of both markets and innovation. As a result, governments have more latitude to intervene proactively as a market takes shape. We will see, for example, that civil society groups in both the United States and Europe worried that biotechnology patents would transform life into a commodity, which would be devalued as it was bought and sold in the marketplace. But while this concern was largely dismissed in the United States, European policymakers took this concern seriously and introduced multiple limitations to patentability as a result.

Consider briefly how these differences have shaped US and European politics related to agricultural biotechnology. Since the field's emergence in the 1970s, citizens in both places have expressed concerns regarding the health, environmental, socioeconomic, and moral implications of this technology.[60] The United States responded to these concerns by developing a limited regulatory framework that would maximize the market potential of agricultural biotechnology, treating genetically modified organisms as "substantially equivalent" to their conventionally bred counterparts. To the extent that it developed regulatory interventions, it focused on acute health and environmental risks that were easily quantifiable and seemingly objective.[61] This kind of intervention, it seemed, would facilitate the rational operation of the marketplace. And while genetically modified foods have become more contested in the United States in recent years, both citizens and policymakers have focused on the question of labeling, thereby placing the ultimate decision-making authority over moral, health, and environmental concerns in the hands of the consumer.[62] Meanwhile, the European Union and individual European countries have generally followed a more proactive and interventionist approach.[63] Reluctant to simply allow the market to operate with minimal oversight, most European countries first identified genetic modification as a novel process.[64] Then they developed regulatory frameworks that required, before the technologies were made widely available, not only extensive testing of product safety but also explicit attention to ethical, socioeconomic, and ecosystem concerns.[65]

PATENTS AND POLITICAL ORDER

Throughout this book, I demonstrate how these market-making and market-shaping approaches have figured in the histories and machinery of the US and European patent systems. This, in turn, has shaped the meaning of patents in the two places: in Europe, patents are treated as moral and socioeconomic objects, while in the United States they are understood as solely techno-legal objects. As early as the seventeenth century, we will see in chapter 1, the English government noticed that patent holders were wielding their exclusive rights aggressively, making daily necessities extremely expensive and provoking citizen riots. It responded by prohibiting patents that would cause "public disorder" in its first patent law, issued in 1623. With this law, the English government treated patents as potentially having far-reaching consequences, for which it was responsible. In doing so, it drew no distinction between patents and the implications of patent-based monopolies, assuming governmental responsibility for both. Eventually, similar public-interest provisions appeared across Europe and became known as *ordre public* clauses because they prohibited patents on inventions that were deemed "contrary to public policy or morality." Such clauses seemed to differentiate between public and private interests in the patent system, and gave government the power to step in to protect the public interest.

By contrast, the earliest US patent laws established patents as legal, technical, and economic entities. A patent bureaucracy, staffed with technically trained examiners, would review and grant patents in accordance with science and the law. The country's founders believed that the predictability of this approach would simultaneously benefit inventors and the public, by encouraging invention and allowing a rational marketplace to flourish. As the system

evolved, patent system officials and organized interests took care to emphasize the limitations of patents. When concerns about the socioeconomic implications of patent-based monopolies emerged in the early twentieth century,
they attributed problems to patent holders rather than to patents themselves.
They suggested further that any government intervention would interfere with
the efficiency and objectivity of the system, which would ultimately damage
the country's strength and growth.

These different approaches to patents in the United States and Europe
seemed to disappear in the face of early twentieth-century harmonization
efforts, but they were only hidden. As the two jurisdictions confronted life-
form patentability, these differences would reveal themselves and be reinterpreted for a new age. In addition, the battles themselves would produce different political orderings.[66] In the face of civil society challenges that placed
unprecedented pressure on their patent systems, the US and Europe would
develop different understandings of appropriate participants and types of
participation, relevant knowledge and expertise, roles and responsibilities of
patent system institutions, and life and its governance.

When the first life-form patent challengers demanded attention to the commodification of nature through life-form patents and invoked ethics expertise in support, traditional participants in the United States argued that such
issues and evidence were irrelevant in a system focused on technical and legal
matters. As the debates evolved, challengers confronted not only rhetoric, but
also bureaucratic practices and rules as well as legal frameworks, that made
it essentially impossible for the domain to consider arguments emphasizing
the moral, distributive, and environmental impacts of patents. This included
a PTO requirement that patents could only be "re-examined" on the basis of
"prior art" (scientific publications and previously granted patents). Similarly,
legal "standing" rules required plaintiffs to demonstrate that they had suffered
direct harm as a result of the "conduct complained of," and that a positive
court decision could directly help them. This implicitly privileged economic
implications and disadvantaged other public-interest claims. As a result, it was
difficult for those inclined to challenge life-form patentability on moral, socioeconomic, or environmental grounds to become a recognized constituency.
Together, these rhetorical, legal, and bureaucratic elements formed what I call
an "expertise barrier": formal and informal rules that develop over the course
of a policy domain's history and are shaped by the jurisdiction's political culture, and which shape a domain's definition of appropriate avenues for participation as well as of relevant knowledge and expertise.[67] In the case of the

US patent system, these expertise barriers reinforced a narrow, techno-legal domain that valued procedural objectivity. It also rejected alternative forms of knowledge and expertise and new constituencies.

These barriers also reinforced a narrow definition of the patent system's role and responsibilities. Officials and traditional organized interests in the US patent system vigorously fought efforts by life-form patent challengers—from farmers to patient advocates—to frame the patent system as a regulatory domain that influenced the control of and access to technologies. Time and again, we will see that PTO officials, patent lawyers, and inventors drew boundaries between the issues raised by challengers and the work performed by the patent system, which they argued was limited, objective, and merely certified inventions to facilitate an efficient marketplace. With this came the idea that moral issues were the responsibility of patent holders and other market players.

Finally, the US patent system developed laws, policies, and practices that identified the fruits of biotechnology that are the object of this book not as life forms at all, but as technologies. This was clear from the outset, in the debates over the status of Ananda Chakrabarty's microorganism in the *Chakrabarty* case that I discuss in chapter 2. Despite Jeremy Rifkin's attempts to define the organism as a form of *life*, US decision makers focused on whether the organism was part of *nature*. This characterization maintained the scientific and legal framing of the domain and rendered concerns about the impacts of commodifying life forms irrelevant. And with this there was little room for bioethical expertise or reasoning. Of course, the United States would struggle with the valuation of life in other policy arenas dedicated to human embryo and cloning research. But to the patent system, which simply certified inventions for market circulation, these matters were irrelevant.

Initially, European patent system institutions responded to the prospect of life form patents quite similarly to their US counterpart. But as individual citizens and civil society groups began to argue that these institutions were not interpreting their moral responsibilities correctly, they reconsidered their approaches. The European Parliament welcomed the new civil society perspectives on the patent system first, incorporating them into its deliberations on a biotechnology patent law, the Directive for the Legal Protection of Biotechnological Inventions (known as the Biotech Patent Directive, or BPD). Subsequently, even the EPO allowed life-form patent challengers to initiate and weigh in on questions of patentability and considered their concerns seriously. Rules that could have posed barriers to challengers, such as those governing who could participate in EPO proceedings and how they could do so, were interpreted to maximize outsider participation. EPO officials began to question whether the bureaucracy's focus on serving inventors' interest always contributed to the public interest. As they rethought their approach, they would accept and begin to contend with public opinion data, ethics expertise, and evidence related to the socioeconomic and environmental implications of patents. However, the bureaucracy would struggle to incorporate these alternative forms of knowledge into its decision making practices. As a result, by the 2000s the US and European patent systems seemed to have somewhat different constituencies, and be attuned to different issues as well.

In this process, European institutions began to characterize the patent system as part of a larger apparatus for governing technology. The EPO, for example, began to see itself as not only a techno-legal bureaucracy but also as a policy organization taking a proactive role in addressing large societal questions including the ethical challenges posed by biotechnology and the global challenges of ensuring access to essential medicines. We will see that it tried to implement this approach by changing both its decision making process and its bureaucratic culture so that its employees would become more aware of their responsibilities to balance public and private interests. But while European officials and organized interests rejected the US's distinction between the techno-legal and the moral and political, they too drew boundaries around the patent system's scope. They tried to do this by distinguishing between the direct and indirect implications of patents, but as we will see this proved to be quite a challenge.

The European patent system envisioned the relationship between patents and life completely differently as well. Because patents represented moral certifications of inventions, they interpreted life-form patents as potentially affecting how citizens valued life. Therefore, as the controversies unfolded, decision makers began to agree that commodifying life through patents could lead to its devaluation. Their presence in the marketplace warranted restriction. Citing concepts of human and animal dignity that had a long legal legacy across Europe, European patent-system institutions restricted patents on human embryonic stem cells and animals, among other things. In doing so, it established itself as a regulatory domain with some responsibility over the governance of ethically fraught biotechnology.

In sum, I argue that these social and political orderings, and the understandings of patents that they presume, are quietly structuring scholarly and policy discussions about the future of patent systems as well as innovation in the United States, Europe, and beyond. Scholars across multiple disciplines, from law to economics, are becoming increasingly engaged with the questions of whether patent systems are "working," how to "fix" them, and what our expectations should be.[68] But what this book reveals is that even these questions themselves are shaped and constrained by our hidden assumptions about patents, patent governance, and the public interest. In order to produce meaningful change, I argue, we must uncover and contend seriously with whether these assumptions are appropriate and how we may want to maintain or change them.

1

DEFINING THE PUBLIC INTEREST IN THE US AND EUROPEAN PATENT SYSTEMS

Patents appear to be governed through a homogeneous international system.[1] While the system is organized by national and regional jurisdiction, the governing institutions, organized interests, and technical and legal experts involved in each look virtually the same. So too do the laws that reward inventors of new technologies with exclusive rights, which operate according to seemingly straightforward criteria: novelty, inventiveness, utility, and sufficient description in the application. These similarities are the result of more than a century of negotiation, from the 1883 Paris Convention to the World Trade Organization's 1994 Trade-Related Intellectual Property Rights (TRIPS) Agreement, which harmonizes rules for patentability, for copyrights that are awarded for authored work, and for trademarks that protect brand names and symbols.[2] In recent years many countries have also signed bilateral and multilateral trade agreements to strengthen these similarities.[3] Governments and inventors argue that together, these treaties form a global regime that makes it easier for inventions to travel, for inventors to reap rewards across borders, and for markets to become transnational.

But this apparent uniformity masks key differences in the legacies, makeup, and dynamics of the world's patent systems. Throughout this chapter I demonstrate that for centuries even the United States and Europe, who have led most of these harmonization efforts, have understood patents and their appropriate governance quite differently. And they have embodied different definitions of the public interest in their patent systems. These approaches are the result of deep but often overlooked differences in political culture and ideology. These differences matter because they help to explain how and why the United States and Europe would respond so differently to life-form patents, and why their political environments began to look so different as the two jurisdictions navigated these controversies. They also call into question the depth of international patent harmonization.

From almost their earliest days, European governments treated patents as moral and socioeconomic objects that could produce monopolies with both positive and negative effects. They were guided by political ideologies that envisioned the marketplace as a part of a larger and preexisting moral order that they had a duty to shape and maintain. Thus, patent-system institutions had a responsibility to protect the public from harms, which included affronts to public morality, inequitable distribution of goods, risks to national security, and eventually infringements of human rights. The United States eschewed this definition of patents in moral terms and emphasized their status as legal and technical objects. The US government's role was simply to set the conditions for the market to flourish, with the assumption that market activity would ultimately produce the public interest. The inventor's interest, in other words, was the public interest. With this came somewhat different stakeholders, rules, and practices, as well as institutional roles and responsibilities. Over the course of the twentieth century, with the rise of economic harmonization efforts across the world and the development of a pan-European patent system, patent systems in the United States and Europe looked increasingly similar. But, the two places still thought about patents, their governance, and the public interest quite differently. As we will see, these differences would eventually reassert themselves and take on new meaning in the life-form patent debates.

PATENTS AS MORAL AND SOCIOECONOMIC OBJECTS

Patent systems emerged first in fifteenth-century Venice and England, as tools to enable the royal courts to create and expand markets. These courts bestowed patents upon entrepreneurs as privileges, allowing them to commercialize a technology exclusively in a particular jurisdiction.[4] In return, the royal courts received revenues from the often-substantial fees they charged the entrepreneurs, while also benefiting economically from additional market activity, technological development, and technology transfer in the jurisdiction.

But by the sixteenth century, these patents had begun to provoke occasional public anger and frustration in England, because some patent holders used their privileges to set extremely high prices for their goods.[5] This caused particular resentment in the case of patents on daily necessities, including salt and oil. Worried that this could lead to a revolt, the English Parliament issued the 1623 Statute of Monopolies to limit the power of patent holders in a variety of ways.[6] It restricted patent length to fourteen years, calculating that this would give inventors the exclusive advantage of training two generations of apprentices to learn to make and use a new technology (previously, the patent term was set by the royal court, and could be essentially infinite).[7]

It required patent holders to make the invention in the country within a limited period of time or risk revocation. The statute also prohibited patents that were contrary to law or caused harm to the state by raising commodity prices, hurting trade, or being generally "inconvenient." Interpreted later as a public-interest clause that prohibited certain categories of inventions based on public policy or morality concerns, at the time the statute referred to inventions that might cause riots or other kinds of public disorder.[8]

With this law, the English government envisioned circumstances in which patents could contravene the public interest. They could, for example, restrict access to some goods and provoke political dissent. And rather than placing the responsibility for such effects on patent holders, it understood patents themselves as having the potential to produce harms. It also understood that it was responsible for stepping in to reduce these harms and to ensure the benefits of patents. To put it bluntly, the government considered itself responsible for shaping the impacts of patents and the market.

More than a century later, the French began to develop what is now known as the "modern" patent system. They reconceptualized patents as bundles of natural legal rights, rather than simply as exclusive privileges, that gave their owners greater power and focused on rewarding innovation rather than just entrepreneurship.[9] France's 1791 patent law also included a "failure to work" provision and an *ordre public* clause that echoed the public-interest language in the Statute of Monopolies. The French understood the *ordre public* clause to prohibit patents on any technology deemed contrary to public policy or morality.[10] Again, the state defined patents as potentially having both beneficial and problematic effects, and it assumed the responsibility of determining when an invention was morally or social problematic and then intervening to ameliorate these problems. But this clause, which would eventually play an important role in the twentieth-century life-form patent debates, was not unique to patent law. An *ordre public* clause would appear in many French administrative laws in the wake of the revolution as a means of demonstrating the "sovereignty of the people."[11] While the meaning of this clause has evolved over the centuries, it has always retained the idea that politicians and administrators should be working on behalf of the public.[12]

This understanding of patents as having multiple — but primarily socioeconomic — impacts on the public interest, potentially hurting citizens, distorting markets, and stifling technological and industrial growth, continued into the nineteenth century. In 1844, the French banned patents on foods and pharmaceuticals, continuing to link patents to the reduced access that might result from patent-based monopolies.[13] It was immoral, they decided, to allow patents that might restrict access to such essential items. This law also strengthened the idea that certain kinds of innovation should be kept separate from market forces. By this time, other European countries had begun to establish patent systems, and many included public-interest clauses, categorical exceptions to patents on daily necessities, and "failure to work" provisions.[14] There were, however, differing interpretations: Italy used the public-interest exception to disallow patents "detrimental to health," for example, while Ireland prohibited "inventions liable to cause an increase in the price of commodities, a hindrance to the freedom of commerce or any other public inconvenience."[15] The British banned patents on "improper sexual appliances" during the Victorian era.[16] And some countries simply banned patents on food and pharmaceuticals, while others prohibited patents on pharmaceutical products but allowed patents on the processes of producing them, hoping that competition and lower prices would result if multiple inventors could market the same product with exclusive rights of commercialization only on the process.[17] While all of these countries saw patents as having moral meaning, they differed in how they interpreted this meaning and in the technologies and patents that they saw as most problematic.

These laws suggest that many European governments were ambivalent about patents in the nineteenth century, and they took on the responsibility of intervening to minimize negative impacts. But things changed in the twentieth century. Many governments became convinced that patents were key to economic growth, particularly because the US and German systems, as I discuss in the next sections, seemed so successful.[18] In addition to participating in the 1883 Paris Convention, some countries began to roll back their patent prohibitions. But there was still concern about the monopolies patents produced. In addition, the two world wars had generated worries that patents could essentially become weapons if foreign inventors used them to restrict the availability of important technologies—including pharmaceuticals—that might be needed on the battlefield.[19]

So European countries developed softer provisions. These included compulsory licensing, which gave governments the power, under specified circumstances, to step in and force patent holders to allow others to make and sell an invention if the patent holder either refused to do so or set the prices of its invention too high. The first such language appeared in Germany's 1911 Patent Law, which allowed compulsory licensing if it was "called for in the public interest."[20] But the language spread quickly after 1925, when the Paris Convention was revised to recognize governments' rights to pass compulsory licensing laws. (The United States did not incorporate such a law, as I discuss later in this chapter.)[21] Over the next few years, most European countries incorporated some version of compulsory licensing in their patent laws.[22] As they did so, many removed blanket prohibitions on food and pharmaceutical patents. They calculated that compulsory licensing laws would give their governments the power and flexibility to step in if these products were not made available at the lowest possible prices, while still empowering patent holders.[23]

In sum, European countries have a long history of conceptualizing patents as innovation and market drivers as well as moral and socioeconomic objects. Although patents could provide necessary incentives for innovation, they could also influence public morality, social and economic welfare, and national security directly. Governments took an active role in trying to minimize these negative effects, while encouraging the patent system's benefits for economic growth. The public's interest, in other words, was in balancing these benefits and harms. Governments also consistently retained the power to shape the marketplace, even by intervening after a patent was issued and used. While many European countries began to think about patents more positively by the early twentieth century, they still had laws on the books that reflected this wariness and their understanding that government had an active role in shaping social life, which included the marketplace.

PATENTS AS TECHNO-LEGAL OBJECTS

At first glance, the patent system envisioned by the founders of the United States looked quite similar to the one that the French built. Like the French, early Americans saw patents as legal rights granted to inventors of new technologies.[24] They articulated these rights in the US Constitution, giving Congress the power "To promote the Progress of Science and useful Arts, by securing for limited Times to Authors and Inventors the exclusive Right to their respective Writings and Discoveries."[25] The founders also seemed to agree that promising patent exclusivity to inventors would stimulate innovation, which would in turn produce and expand markets. But these exclusive rights only lasted for a limited period of time, to ensure that others could build upon the invention and develop new, potentially patentable, technologies themselves.

But as the system developed, it became clear that the United States understood patents somewhat differently than its European counterparts. The patent laws that gave the system its shape in the late eighteenth and early nineteenth centuries did not articulate any explicit considerations or exceptions on the basis of the public interest; rather, they envisioned patents as technical objects that drove innovation and markets but had minimal—if any—impacts on public morality or access to technology. They emphasized objectivity and transparency in patent-system procedures and the importance of technical expertise in reviewing applications. As policy controversies emerged and were resolved through the nineteenth century and into the twentieth, the US patent system seemed to settle on the idea that the public interest would be best served if the patent system and the market were left alone to work, with government playing a minimal, certifying role.

A Scientific Approach to Patent Decision Making

Passed by the US Congress in 1790, the first patent law provides early clues about this orientation.[26] Creating rules for how the system should function, it articulated for the first time an examination system. Most contemporary European systems required only that inventors register their patents with the government, but the 1790 law required the US secretaries of state and war, and the attorney general, to review the utility and importance of inventions. The new court system would adjudicate disputes over these decisions or over a patent's novelty.

The law required, also for the first time in any national law, that applicants submit a written "specification" and a "draft or model." These submissions were necessary "to enable a workman or other person skilled in the art of manufacture . . . to make, construct, or use the same, to the end that the public may have the full benefit thereof, after the expiration of the patent term."[27] Early administrators of the patent system then put some of these drafts and models of inventions on public display, to promote patents as points of civic pride. They were part of efforts to promote "industrial tourism" and to increase the production of "useful knowledge."[28] Historian Mario Biagioli and others have argued that the public availability of the written specifications and models, coupled with low application fees, invited a much larger group of potential inventors than did the intellectual property systems that existed in Europe at the time.[29] Indeed, the patent system, the inventions themselves, and the details regarding these inventions were more accessible than ever before. These scholars suggest that this made the innovation process more democratic. Viewed differently, however, these elements assumed that the public was composed of either inventors or potential inventors, ready for and interested in engaging in the innovation process. This approach also assumed that the interests of inventors and non-inventors were the same, and that they would all benefit from a strong patent regime that produced more technologies for the marketplace. This was, in essence, the moral basis of the US patent system.

The 1836 Patent Act, which established the basic institutions and rules for the system that the United States knows today, reinforced this early approach. Importantly, it defined patents as techno-legal objects to be governed with procedural objectivity. With this law, Congress created a central patent bureaucracy (then the Patents Office, now known as the Patent and Trademark Office, PTO) that would review applications and grant patents. The law authorized an examining clerk and two assistant clerks to review applications. One of these assistant clerks had to be a machinist and the other a "competent draughtsman" who possessed both technical knowledge and the skill to produce engineering drawings. Both had to have the expertise to ensure that all granted patents were "new and useful" inventions.[30] Patent applicants could appeal these examiners' decisions to an internal appeals board, made up of experienced machinists and draftsmen (known today as the Patent Trial and Appeal Board) and ultimately to the courts. A commissioner of patents, appointed by the president of the United States, would oversee the bureaucracy's work. But the courts would play a central role in negotiating patent disputes, interpreting patent law, and developing case precedent.

This law also further articulated the written disclosure requirement. The previous law had required that a description of the invention should be comprehensible to "a workman or other person skilled in the art or manufacture," but it now focused on comprehension by "any person skilled in the art or science to which it appertains."[31] This language, coupled with the expertise desired of examiners, underscored the idea that technically trained personnel were the appropriate authorities on patent-related matters and would confer legitimacy on the new patent system.

The standards that early Patent Office examiners had to meet were extremely high because they were seen as doing vital work for the government and the new nation. Early congressmen argued that the patent clerks needed a unique blend of technical skill, scientific and legal knowledge, and competency in the French language (because French science and technology were well developed at this time). John Ruggles, chairman of the Senate Committee on Patents, described an examiner's ideal qualifications in 1836: "An efficient and just discharge of the duties, it is obvious, requires extensive scientific attainments, and a general knowledge of the arts, manufactures, and the mechanism used in every branch of business in which improvements are sought to be patented. . . . He must moreover possess a familiar knowledge of the statute and common law on the subject, and the judicial decisions both in England and our own country, in patent cases."[32]

Indeed, the first examiners employed by the Patent Office seemed to possess the characteristics that Ruggles outlined. All had a deep understanding of natural philosophy and specific scientific fields (often with advanced formal training); some had hands-on technical ability (draftsmanship, for example); some had formal legal training as well.[33] All seemed to be highly respected. In the 1840s, the commissioner of patents referred to them as a "living encyclopedia of science," while the new periodical *Scientific American* (owned by a patent agent, it had begun to represent the perspectives of inventors and the patent agents who had technical backgrounds and helped them file applications) argued that their duties were "more arduous than an Ambassador's or a Cabinet Minister's."[34] This examination approach suggested that the judgments of examiners should be trusted because they had multiple and diverse skill sets and an extraordinary level of knowledge, and because the work was extremely difficult. Furthermore, because they based their judgments on this vast knowledge, they were thought unlikely to be biased or incorrect. The American emphasis on scientific examination was quite different from what was emerging in Europe at the time. Prerevolutionary France had had an informal examination system, in which members of its Academy of Sciences weighed in on the government's patent decisions.[35] Its 1791 law, however, eschewed this approach and required inventors simply to register their patents.[36] Other European governments also had registration systems, largely to promote the system's efficiency: inventors would not have to worry about lengthy and uncertain examination processes.

Even in the United States, however, patent examination was not straightforward. It required some negotiation with the patent system's emerging stakeholder community.[37] The new bureaucrats were accustomed to the standards for demonstrating novelty, discovery, and proof in their scientific fields. As a result, they initially rejected many of the patent applications they saw, deciding that they described obvious technologies that were therefore unpatentable in the context of known theories.[38] This was deeply distressing to prospective inventors, patent agents, and patent lawyers, who argued that they would not be able to create markets if the patent requirements were so stringent. They were more comfortable with an approach closer to the European registration systems.[39] This conflict, as Kara Swanson has observed, led to years of negotiation.[40] But by the 1860s, the technically trained examiners had loosened their requirements somewhat. As a consequence, patent-system stakeholders became more accepting of understanding patents as technical objects and of the relevance of technical expertise in the patent system.[41] The definition of the patent evolved, in other words, to gain the acceptance of system stakeholders.

Over the next decades and into the twentieth century, the US patent system grew dramatically, and its technical orientation continued. More and more people applied for patents,[42] and as a result, the Patent Office hired more examiners. They no longer met the exacting standards of the early congressmen, but all held at least a bachelor's degree in the natural or physical sciences or engineering.[43] With this growth also came more systematic and specialized processes. The Patent Office divided its personnel into sections based on their expertise, and developed systematic processes for evaluating patent applications as they traveled through the bureaucracy. They also developed specialized training (including a manual) for examiners, so that they could easily access the relevant scientific, engineering, and legal knowledge, which might ensure objectivity and reliability in decision making. Yaron Ezrahi and Sheila Jasanoff have both noted that procedural objectivity in bureaucratic decision making is an important part of American political culture.[44] There is little trust in bureaucratic or expert judgment; instead, trust in bureaucratic decisions comes from the belief that bureaucrats are making decisions solely on the basis of clear evidence. The patent bureaucracy followed this tradition as well, basing judgments on previous patents and scientific publications (called "prior art") and legal rules. Personnel placed (and continue to place) great emphasis on demonstrating procedural objectivity through adherence to scientific evidence as well as clear rules and procedures.[45]

Resisting Alternative Understandings

The emerging approach to patents in the United States faced repeated challenges, particularly during the world wars and as Europeans passed compulsory licensing laws. But time and again, US decision makers and organized interests defined patents and the patent system as focused solely on stimulating innovation. Most other public concerns were seen as minor.

One of the first challenges came in 1817, when the Supreme Court heard the *Lowell v. Lewis* case. Both parties in the case owned patents on water pumps, and Lowell argued not only that Lewis infringed his patent, but also that because a water pump was already on the market, his invention was not useful and did not deserve a patent. The court found that both patents were valid, and that the utility criterion for patentability did not require that an invention be better, only different. But the court ruling also articulated what came to be known as a "moral utility" doctrine. In his opinion, Supreme Court Justice Joseph Storey suggested that there might be rare occasions when patents could be prohibited because they did not have a beneficial use, but rather were pernicious, frivolous, or worthless. Scholars have interpreted this doctrine as an analog to the public interest and *ordre public* clauses common in Europe.[46] But it was drawn more narrowly than the European analogs that had developed by then: "A new invention to poison people, or to promote debauchery, or to facilitate private assassination, is not a patentable invention. But if the invention steers wide of these objections, whether it be more or less useful is a circumstance very material to the interests of the patentee, but *of no importance to the public* [emphasis added]."[47] Except in extreme circumstances, the use of the invention was deemed an irrelevant consideration. After all, the public's interest was in making more technologies available in the marketplace. Of course, later courts could have interpreted the moral utility doctrine broadly. But instead, the utility criterion would get weaker and weaker in US patent law.[48] By the twentieth century, the moral utility doctrine appeared extremely rarely in American jurisprudence.[49] Bureaucrats, the courts, policymakers, and even traditional interests seemed to be extremely reluctant to envision patents as having harmful effects or to test the moral utility doctrine's scope.[50] This suggested that they were reticent to take a more active role in shaping the market.

In the mid-nineteenth century, medicinal patents posed another set of challenges to the US patent system's understanding of its role and of the public interest. As historian Joseph Gabriel has written, pharmaceuticals were becoming increasingly popular among the public as convenient remedies to their ailments, and despite some concerns about their effectiveness (they were not subject to regulatory control), they were usually patented and trademarked.[51] Physicians were quite concerned about this development, being skeptical about the utility of pharmaceuticals in comparison to their own hands-on medical practice. In particular, they opposed the introduction of financial incentives into medicine through these patented drugs, worried that patents would interfere with the development of a common storehouse of scientific knowledge—and therefore medicine.

What is particularly important, in comparative perspective, is what happened next. As discussed previously, many European countries banned patents on pharmaceuticals or at least on the products (allowing patents on the processes) around this time. This assumed government responsibility regarding these issues. But in the United States, the government never banned pharmaceutical patents. Instead, the professional association representing potential market players (the American Medical Association, or AMA) took on the responsibility of guiding the behavior of potential patent holders and issued a code of ethics prohibiting physicians from owning patents. But as many scholars have observed, there was a vigorous market in "patent medicines" in the nineteenth century. By the end of the century, physicians eschewed the AMA guidelines and applied for patents in greater numbers, and pharmaceuticals became more reliable.[52] Whereas pharmaceutical patents and their potential implications were seen as the government's and specifically the patent system's responsibility across Europe, these responsibilities were left to market players in the United States.

The concern that patents might interfere with common medical knowledge and practice, and access to technologies more generally, emerged again in the early twentieth century. When European countries turned to compulsory licensing, the United States also considered it. But the United States approached it more narrowly and ultimately came to a different conclusion regarding its relevance and viability. In 1912, Representative William Oldfield, a Democrat from Arkansas who chaired the House of Representatives' Committee on Patents, initiated hearings to consider a "failure to work" provision in the patent laws.[53] In 1910, the US Congress had enacted a law that allowed a patentee to sue the government for compensation if it used his invention without a license, but this law did not provide the government with the explicit authority to license inventions on a compulsory basis.[54] Oldfield defended his failure-to-work provision by arguing that it was much less intrusive and more pro-innovation than the similar laws adopted by European countries: "Nearly all countries except the United States have provisions of law requiring the working of a patented invention, and in the event of failure to adequately work the invention that the patent shall either be revoked or that the owner thereof shall be required to grant a license to others to manufacture, use, and sell the same. In my opinion the so-called working clauses contained in the laws of other countries are of such a drastic nature as to discourage invention."[55]

Oldfield's amendment did not challenge the idea that patents drove the market and innovation on a broad scale. And in fact, he was not even considering an explicit "public interest" provision that many European countries had or would adopt. But he was suggesting that patents could, on occasion, hurt the availability of technologies. And he was envisioning a more interventionist approach for government, assuming that patent holders may not always know best (or operate in ways that benefitted the public interest). The patent profession and manufacturers responded vigorously and negatively. While the House Committee granted 27 days of hearings over the course of 5 months to hear their opposition, nobody spoke in support except for Congressmen.

Testimony in these Oldfield hearings reinforces our emerging picture of how patents, their governance, and the public interest were understood in the United States. The vast majority of witnesses were patent lawyers, but a few individual inventors and representatives of companies spoke as well.[56] They all agreed that the legislation would destroy the patent system, as well as innovation and economic growth in the United States. Without patents, they argued, society would lose because it would have less access to innovation and fewer markets. William Dodge (who represented the Patent Law Association and would become a frequent Congressional witness on behalf of both patent lawyers and manufacturers) stated for example: "The constitutional provision has encouraged [the inventor] to go ahead and perfect the thing, and he does it because of that encouragement and in the hope of the protection which it holds out to him, and without which encouragement you will never get this advance."[57] Dodge argued further that suppression of patented inventions was simply not a problem, and certainly did not warrant what he considered to be a significant policy change.

> I find from my experience in connection with these manufacturing concerns, and I may say it is a long one — longer than you might imagine — that the tendency is to put out the best machine they can produce as promptly as they can try it out, and know that it will fulfill its requirements, and that they do almost universally reduce the price in so doing. I know that to be so in various lines of manufacture of machines that are used all over this country and all over the world. It is not a fact that they suppress these inventions.[58]

This perspective was notable for two reasons. First, it underscored the idea that patents were innovation and market drivers, and Dodge declined to acknowledge variability in these effects. Rational market participants did not suppress their patented inventions, and thus the government could rely on them to behave appropriately. Second, Dodge established industrialists and their legal representatives as the appropriate experts because they brought direct experience with the patent system. Over the next decades, as others tried to establish their expertise as relevant, their lack of industrial experience would make it difficult for them to do so.

The legislation suggested by Representative Oldfield failed. But the issue emerged again during and after World War I. In 1917, the secretaries of war and of the navy, as well as leaders of the American Medical Association (AMA), became concerned about the limited availability of Salvarsan, a drug used to treat syphilis. The German company Hoechst held US patents on the drug and controlled its use, making it very expensive at a time when syphilis was growing among American military men and the public as a whole.[59] The Senate Committee on Patents held hearings on the matter, and invited physicians, including professors from the Mayo Clinic and nearby Johns Hopkins University, to testify. As they advocated compulsory licensing of Salvarsan, these witnesses questioned the harms the patent system might cause to public health, echoing some of the concerns that had emerged in the nineteenth century. Representatives of the Mayo Clinic argued, "We respectfully submit that even in time of peace the propriety and wisdom of a public policy which permits private monopoly to control and seriously limit the use of remedial agents whose availability is vital to the public health may well be questioned. . . . If it [is] unwise to permit commercial control of the public health in time of peace, it is doubly unwise to permit it at a time when the country is bending its energies to prepare for and successfully wage war."[60] An editorial in the *Journal of the American Medical Association* echoed this sentiment, noting that this was "an instance in which property rights are held higher than fundamental human rights."[61] This was a remarkable argument in the context of the US patent system. Property rights *were* usually described as "fundamental" human rights. Human rights and inventor's rights and interests had not been distinguished in this way before. In addition, physicians had now asserted their expertise in the patent system, in which they had played little role previously.

Congress solved the immediate problem by passing the Trading with the Enemy Act (TWEA). The act set up a government-sponsored trust responsible for the seizure, administration, and sometimes sale of "alien property" during wartime, including the German patent covering Salvarsan.[62] It did not, however, address the broader issue of compulsory licensing in the public interest. In fact, the Congressional hearings regarding Salvarsan did not show much attention to the concerns raised by the medical profession regarding the potential impacts of patents on public health.[63]

The limited scope of the TWEA became clear soon after World War I ended. In 1922, Democratic Senator Augustus O. Stanley, from Kentucky, was concerned that the Germans had received US patents on technologies related to national security and then used these patents to stop development and commercialization by US entities. So, he introduced compulsory licensing legislation again. Once again, the Senate Committee on Patents held hearings focused on whether patentability should be limited if national security was at stake. Most of the witnesses were the patent system's traditional participants, including independent patent lawyers, the Commissioner of the Patent Office, representatives of the Patent Law Associations of New York, Chicago, and Pittsburgh, and envoys from industrial groups, including the American Institute of Chemical Engineers, the Manufacturing Chemists Association, the Synthetic Organic Chemical Manufacturers Association, and the Federated American Engineering Societies. This time, however, the Judge Advocate General from the War Department also testified, expressing concern about the situation:

"We are up against it, because we have not the manufacturing technique and the manipulation necessary, even though we have the patent and their full disclosures, if we do not have the working of those patents in the United States—that is, the industries throughout—why, we lack a very essential part of our national defense, outside of the economic questions."[64]

But opponents of the legislation responded that any compulsory licensing legislation would hurt innovation. William Dodge appeared again and argued that the best way for the patent system to serve national security was for it to continue to be the engine of America's progress. He noted, "I think everybody will concede it—that any restriction or any provision requiring the working in the country or compulsory licenses in the event that the patentee or assignee fail to work here would tend to lessen the taking of patents in this country on things that could be kept secret."[65] While he acknowledged that the Germans might occasionally take US patents and then use them to block certain areas of innovation, he argued that such circumstances were rare and only lasted for the term of the patent. Intervention in the patent system through compulsory licensing, he argued, would diminish the power of patents as market drivers, because patent holders and competitors would live in fear of government interference. As a result, market rationality would not function properly.

By characterizing the patent system as the "best incentive," Dodge framed his position in a way that would be echoed in the late twentieth- and early twenty-first-century debates focused on life-form patents. In his view, altering the system (in this case through compulsory licensing provisions) posed uncertainties for innovation and the economy. Compared to these uncertainties, he and his allies saw the monopoly and access problems as insignificant and not of concern to overseers. In addition, he linked the prevailing understanding of the public interest in the US patent system—that the public would benefit most if the patent system were allowed to stimulate invention without interference—with an approach to knowledge and expertise that privileged the contributions of the inventor and patent law communities. This contrasted with a view of the public interest that took national security into account and that would have included military officials and perhaps others as experts. Dodge's approach was successful, and opponents of compulsory licensing managed to prevent any new legislation from passing.

For the next few decades, although the rates of patent applications and granted patents remained high, and the US pharmaceutical industry began to flourish, excitement over the patent system in the United States waned. Politicians and the courts began to worry that patents, if too strongly enforced, could actually interfere with the free operation of the market.[66] This led to a handful of court decisions limiting the rights of the inventor.[67] But even during this period, there was great reluctance to consider more systematic government intervention in the patent system. The extensive Kefauver hearings, which have since become well-known because they led to the transformation of pharmaceutical regulation in the United States,[68] illustrate this point. They also demonstrated further the strategies used by insiders to reject alternative understandings of the US patent system's role and responsibility.

Held between 1959 and 1963, these hearings investigated the efficacy of FDA regulations as well as drug prices, which Senator Estes Kefauver, a Democrat from Tennessee, called "excessive and unreasonable."[69] This investigation of the pharmaceutical industry explored a wide range of issues, from drug advertising to anticompetitive practices. The committee also considered the effects of patents on increased drug prices, and the possibility of compulsory licensing emerged again. Like a few congressmen before him, Kefauver and his allies saw patents as socioeconomic objects that could have negative impacts on public health. Given this, it is notable that Kefauver was himself an outsider to the patent system: he didn't sit on the Senate's Patent Subcommittee and instead called the hearings related to drug patents through the Senate Antitrust and Monopoly Subcommittee, which he chaired. Perhaps as a result, he invited both customary participants in the patent system and those who had played less of a role in it in the past, including economists, physicians, and consumer groups, to testify.

In 1961, Kefauver introduced legislation that would require pharmaceutical companies to license their drug patents after three years of market exclusivity, mandate that the names of generic drugs be displayed on labels and in advertising, require federal licensing of drug firms, and increase requirements for drug safety and efficacy.[70] Although President Kennedy supported it, the bill became quite controversial in Congress.[71] Patent lawyers were among the most vigorous opponents, and their approach seemed more strident than before—perhaps due to the US pharmaceutical industry's success at the time.[72] They reiterated the focus on patents as solely innovation drivers: Joseph Jackson, chairman of the American Bar Association's section of patent, trademark, and copyright law, stated, "I believe that the value of patent rights is demonstrated by the wonderfully creative record of the US pharmaceutical industry and especially by the availability to the American public of drugs far superior to those of only 5 years or 10 years ago. It would be penny-wise and pound foolish to enact legislation which would jeopardize the continuous outpouring of these new drugs."[73] But Jackson and his allies also stated clearly, as never before, that socioeconomic concerns related to drug prices were not patent issues. Rather, these issues required other kinds of legislative interventions. Jackson argued that, in his opinion, "if any evils are found to exist in the drug field or any other field, and these evils cannot be attacked by present laws, then legislation directed specifically to overcome these evils should be enacted. Any legislation that does not require a showing that an alleged evil exists should be scrutinized very carefully before being enacted. It would be disastrous if the Congress, in any effort to combat an alleged evil, destroyed an important part of the American patent system."[74]

These arguments defined both patents and the responsibilities of system participants as extremely limited. Patents stimulated technologies and markets, and did not hurt public health in terms of the cost, effectiveness, or availability of drugs. Any harms (or, to put it in Jackson's terms, "evils") were the responsibility of other areas of government, not the patent system. Furthermore, the "continuous outpouring" of new drugs demonstrated that it was in the patent holder's best interest to make his or her technology widely available in the marketplace. In other words, she had an interest in behaving morally in order to make money. Within this view, the responsibility of the government in the context of the patent system was only to ensure that patent decisions were made objectively.

Given the focus on markets, we might assume that American economists would play an important role in these debates. But, they did not. This was partially because few economists had focused on patents but also because those who did at the time saw them as interfering in free markets.[75] The situation changed with the Kefauver hearings, when Fritz Machlup, a well-known economist who would later become the president of both the American Economic Association and the International Economic Association, testified. He had written extensively on the patent system, even performing an "economic review" of it for Congress in 1958.[76] In the hearing, he expressed support for compulsory licensing of drugs for both moral and economic reasons, arguing that drug patents were too harmful to public health and that other incentives—including the possibility of fame and the satisfaction that came from a pivotal discovery—could be enough to drive pharmaceutical innovation.[77] But Congressmen repeatedly rejected his ideas as ignorant and biased. His interchange with Republican Senator Roman Hruska, from Nebraska, is representative of this engagement:

> MACHLUP: A few weeks ago I received a letter from a patent lawyer
> who asked, 'If it was not the patent system that has made America
> great, what else was it?' So I wrote back to him and said if he polled

a hundred economists and asked them what has made America great, they may list 10, 15, 20 different things. I doubt that the patent system would be among them. The patent system may have been good, the patent system may have been bad, I do not know. But I do not think that any economist—and I mean recognized economists—would hold that the patent system was one of the major forces. It may have been a good thing. It may have promoted the arts and sciences, or it may not have; we cannot know.

HRUSKA: But you know, Professor, the economists for over a hundred years have been telling that story. They have had a bias against patents. The literature proves that. . . . Yet the composite judgment of businessmen, of parliamentarians, of the financiers, of those in charge of the Nation's affairs, has gone the other way and is going the other way. . . .

MACHLUP: What you call the composite judgment I call the increased influence of pressure groups. It is true that economists have been, by and large, skeptical about the patent system. It is also true that after 1873 they were almost not heard on the patent system. If you once check the hearings on patent laws here in the US Congress, you will notice that until about the late 1930s, hardly any economist was ever heard on the patent system.[78]

Hruska insinuated that while Machlup might have knowledge of economic theories related to patents, his lack of experiential expertise and economists' bias against the patent system disqualified him from participation. He and other insiders refused to engage with the substance of Machlup's arguments. This is not surprising. Engaging with Machlup would have acknowledged the argument's legitimacy. Instead, by challenging the economist's credibility and expertise and emphasizing his bias, they could argue that this perspective did not belong in the patent system. Machlup was neither knowledgeable in the appropriate ways nor objective. While economists provided occasional testimony in Congressional discussion (particularly as debate about the government's patent policies became more intense through the 1960s and 1970s), they were rare and peripheral figures in the US patent system even through the twentieth century, and the patent system was not a major area of research for economists until relatively recently:[79] the PTO appointed its first chief economist only in 2010. This is perhaps surprising, given the important role that economists play in many other areas of public policy in the United States, but it makes sense, given their historical skepticism.[80]

Representatives of consumer groups who testified at the Kefauver hearings, who were also critical of drug patents, suffered a similar fate. Colston Warne, the president of the Consumers Union, publisher of the magazine *Consumer Reports*, testified in favor of the approach taken in many European countries at the time, limiting patents to the process of making drugs. He justified his position, "In modern life, in the crowded living conditions of modern society where the health of us all is so closely interwoven with the health of a given individual, the drugs by which we control disease and feud against death cannot be viewed simply as products of an economic striving." [81] "Are you against profits, Mr. Warne?" Hruska responded. With this bold question, Hruska wasn't simply suggesting that profits were the primary goal in the patent system, and that improvements to public health would follow. He was also characterizing the government's role as supporting the companies that produced the pharmaceuticals. This approach, of course, fit generally with the way that the United States had understood the relationship between the government and markets. The government was not to interfere beyond granting patents to deserving inventions.

The legislation that the Kefauver hearings ultimately produced, and which was signed into law, extended the Food and Drug Administration's powers in terms of drug safety but was silent on drug prices and patents.[82] Kefauver and his supporters had not managed to redefine the patent system as a regulatory space that influenced drug prices and therefore access.

By the 1970s, then, the United States had established patents as limited techno-legal objects that needed no additional government involvement beyond certifying inventions. On multiple occasions, decision makers had dismissed the idea that patents encouraged monopolies that hurt society. Overall, they repeated, patents and inventors were benefiting society by increasing innovation and encouraging markets. In fact, Congress had intervened to limit patentability extremely rarely, and only in cases related to national security. Its Trading with the Enemy Act had dealt with foreign patent holders during wartime, while legislation passed during the 1940s and 1950s maintained strict controls on who could own patents related to atomic weaponry.[83] National security, it seemed, was the only concern that could penetrate the otherwise seemingly amoral world of the US patent system.

BUILDING A PAN-EUROPEAN SYSTEM

As the United States developed and reinforced its distinctive approach to patents and their governance throughout the nineteenth century and into the twentieth, there was still considerable variation across the European patent systems. Most had legal registration systems, but in 1877 Germany adopted a technical examination approach that echoed the one in the United States.[84] Patentability rules differed as well.[85] One of the few areas of similarity was "public-interest" language in the patent laws, which of course was absent in the United States but did not seem to figure much in the day-to-day operations of these European patent systems.[86]

But slowly, encouraged by international harmonization efforts, the European systems began to remake themselves to look like one another and like their US counterpart. Concluding that the US and German systems had produced both innovation and economic growth, other European countries incorporated technical examination into their patent procedures.[87] In 1941, Belgium, the Netherlands, and Luxembourg (known as the "Benelux" countries) established the International Patent Institute (IPI) in The Hague to consolidate efforts to determine a patent's novelty.[88] Technical personnel drawn from the three countries would investigate scientific publications and previous patents and produce a "search report," which was then used by national patent offices in the examination process. But as we shall see, even though European patent systems began to look like the one in the United States, they retained elements of their distinctive approach to the public interest, which suggested that much bigger differences still lay underneath.

One of the most comprehensive harmonization efforts began after World War II. In 1949, the Council of Europe (the Council), made up of Belgium, Denmark, France, Ireland, Italy, Luxembourg, the Netherlands, Norway, Sweden, and the United Kingdom, initiated discussion of a pan-European system in its Committee on Legal and Administrative Questions. The Council hoped that a unified patent system could "achieve a greater unity between its members for . . . facilitating their economic and social progress.".[89] The committee, which was responsible for all of the Council's legal and human rights matters, first proposed a system much like the one in the United States, defining patents as techno-legal objects. A European Patents Office composed of technical experts would ascertain, "through cross-questioning if necessary," the "novelty and patentability" of an invention that was necessary to "technical progress."[90] A system based on these technical procedures would produce objective decisions, they hoped, which would facilitate trade across countries and, as a result, strengthen European peace.

While the Council largely agreed with this proposal, it decided that it was not yet "practicable" to implement. It would require an unprecedented level of political harmonization, raising questions about the fate of national patent offices and the degree of influence each country would have in a pan-European system. Negotiators therefore focused first on negotiating the processes for patent examination and the substantive aspects of patent law, including developing common definitions of "industrial character," "novelty," "technical progress and creative effort," "scopes and roles of the description and the claims," and *ordre public*.[91] They convened a Committee of Experts on Patents, made up of national representatives of patent offices, to do this work. While these seem like technical processes, they were also deeply political and would influence the administration of the patent system.

Defining Patentability

Although negotiations over the rules for patentability would take twelve years, European countries agreed quickly to include an *ordre public* clause.[92] Because it had been used rarely,[93] it may have seemed unimportant, particularly in the context of the other issues up for debate. There was some discussion, however, of how the clause should be interpreted. For example, the committee chair argued that legislators should have the final say because they were representatives of the people, but the Dutch representative worried that this might lead to an expansive definition and limit patentability too severely.[94] The final agreement would not include much guidance for interpretation.

There was much more conflict over whether a pan-European system should exclude certain types of inventions from patentability. Some countries still had different laws about the patentability of "chemical products," including pharmaceuticals, "foodstuffs," and "horticultural products."[95] In the case of pharmaceuticals, some allowed patents on both the products and the processes of making them. But most either prohibited "product" patents because they facilitated monopolies that would be detrimental to the public's health or allowed patents on products but also allowed compulsory licensing under certain circumstances in order to deal with the potential problems that a monopoly could create.[96] The Austrian delegation described its position as follows: "For the Austrian economy, the protection of chemical substances is unacceptable. Furthermore, it seems entirely inadmissible that foodstuffs, stimulants, pharmaceuticals, and disinfectants necessary for the preservation of the life and health of the general public should be subject to exclusive rights for the benefit of individuals."[97] It saw patents as being connected directly to questions of life, health, and public needs. The responsibility of the government, and the patent system in particular, was to balance these concerns against the inventor's rights in order to benefit society.

Industry representatives and patent-law organizations also weighed in on this issue.[98] Notably, particularly in comparison to the US debate, they seemed to accept, at least publicly, the relevance of social welfare concerns. But they argued that categorical exceptions to patentability provided blunt instruments with which to achieve the public interest, particularly given existing compulsory licensing laws. In 1962, for example, the Committee of National Institutes of Patent Agents, an independent organization that represented patent agents in multiple European countries, noted the following in relation to the patentability of foods:

> With special regard to the stipulation excluding foods, it is emphasized that they were incorporated in the patent laws at a time when the number of commonly available foods were far more limited than now. Under present conditions it is difficult to believe that a new food will be invented which will have such decisive influence on social conditions and nourishment that the production therefore immediately can be started by everybody. Should the patent be exploited to the detriment of society, there are now far better possibilities than previously to remedy the situation by means of monopoly and price controls. Also compulsory licensing will offer the opportunity to create sound competition.[99]

Similar interests in the United States had argued that patents simply stimulated innovation and that any monopolistic effects were the responsibility of the patent holder or other policy domains, not the patent itself. By contrast, in Europe these interests accepted the relevance of social concerns and monopolistic effects as well as the authority of government to step in and shape the marketplace. But they argued that compulsory licensing legislation provided sufficient authority for government to operate in the public interest—categorical exceptions were unnecessary. In the end, all Council countries signed what became known as the Strasbourg Convention on the Unification of Certain Points of Substantive Law on Patents for Invention (the Strasbourg Convention) in 1963. It allowed patents on food and pharmaceutical products, as well as on agricultural and horticultural processes. In other words, they erred on the side of expanding patentability.

But they did not drop their public-interest concerns entirely. They eliminated the categorical exclusions because negotiators seemed to agree—after much discussion—that compulsory licensing legislation at the national level would suffice to ensure that food and medicine were available "to the public at the lower prices consistent with a reasonable advantage to the patentees."[100] The law also included an *ordre public* clause prohibiting "inventions the publication or exploitation of which would be contrary to *ordre public* or morality, provided that the exploitation shall not be deemed to be so contrary merely because it is prohibited by a law or regulation."[101] And it also prohibited patents on "plant or animal varieties." Europe already had a plant-variety protection system, which it had created to comply with the 1961 International Convention for the Protection of New Varieties of Plants (UPOV).[102] UPOV, which the United States had also signed, enabled intellectual property protections for farmers and breeders in order to encourage the development of new agricultural varieties. But the plant-variety protections were slightly weaker than the rights provided by the "utility" patents, which were the focus of the Strasbourg Convention and are the focus of this book. For example, plant-variety protections explicitly allow farmers to save seed across generations, while utility patents have no such automatic exclusion.[103]

The 1973 European Patent Convention (EPC) that finally established the pan-European system would rely heavily on this Strasbourg Convention. The only addition was a third exclusion covering "methods of treatment of the human or animal body by surgery or therapy and diagnostic methods practised on the human or animal body."[104] This provision, which appeared in patent laws across Europe, was designed to eliminate conflicts of interest for health care providers.[105] If these providers were simply prohibited from patenting their diagnostic or therapeutic methods, the thinking went, then no ethical conflict would arise. This prohibition again demonstrated a very different understanding of the role of government in comparison to the United States. US government decision makers had generally repudiated the idea that patents created a conflict of interest, emphasizing their limited scope and the moral and rational actions of patent holders. They left such questions up to physicians themselves. But EPC negotiators and national governments across Europe envisioned patents quite differently and assumed a more active role for government in regulating the morality of both patents and market players who were, in this case, physicians.

Defining Governance

Although the Strasbourg Convention provided some guidance for the governance of a pan-European patent system, there were still many other questions to be resolved. Would patents be European, and if so, what would happen to the national infrastructures that individual countries had built? How would power over a new European system be allocated? These were not simply logistical questions, given the centuries of conflict between some of these countries as well as nationalist sentiments. As one of the earliest efforts in pan-European governance, the new system could have significant implications for the shape not only of the European market but of European policymaking as well.

As countries debated the Strasbourg Convention, two groupings of European trading partners offered proposals on these matters. The European Economic Community (EEC), created in 1957 among Italy, Germany, France, and the Benelux countries, proposed a pan-European approach that maintained the sovereignty of the national systems.[106] It proposed a European Patent Law that would establish a central European Patent Office to perform search, examination, and granting functions, with the European Court of Justice and an Administrative Council made up of representatives from the EEC member states overseeing its activities. The new office would operate in parallel with the national offices. Non-EEC countries could participate if the Administrative Council approved them.

This proposal envisioned both national and European patents, with the structure of the European system echoing its American counterpart with a central bureaucratic apparatus and external legislative and judicial oversight. The pan-European system would be attractive to those who sought to market their products across national borders. However, this system would create a new layer of pan-European decision-making authority and expertise, which would coexist with the expertise and authority of national systems. Thus, it would create new challenges. What would European expertise in patent searching and examination mean? Would member countries have equal representation among patent examiners? Would non-EEC countries accept decisions by the Administrative Council when they were not represented in it? And, perhaps most important, would there be an effort to harmonize European and national approaches to patentability?

Because EEC negotiators could not agree on whether and how non-EEC countries could participate in the proposed system, their draft proposal did not make much progress. In the meantime, the heads of patent offices from the member states of the European Free Trade Association (made up of Austria, Denmark, Norway, Sweden, Switzerland, and the United Kingdom), proposed a "two-part arrangement" in which there would be a unified system up to and including the grant of the patent, but then policies regarding the effects and validity of the granted patent would be confined to individual countries. National offices would grant patents as well. Under this arrangement, a prospective inventor who wanted to market her technologies across Europe would be more likely to use the European system, because she could ask the European office to consider her application for multiple national patents. But if she only wanted patent protection in one or two countries, she would be more likely to use the national patent offices. In this proposal, a European office would have fewer responsibilities, and member countries would maintain more autonomy. This would resonate with and help to solidify the emerging concept of a unified Europe, which had a growing social, political, and economic meaning, but also maintained a clear space for national autonomy.[107]

The EFTA Memorandum attracted a great deal of interest. The French government, primarily at the request of French industry, advocated this proposal inside the EEC, and soon the chairman of the EEC's expert committee incorporated it into a draft memorandum.[108] It soon became the basis for discussion at the EEC's Inter-Governmental Conference, which was made up of the directors of the patent offices of its own member states and included observers from the Council of Europe, the International Patent Institute, the United International Bureaux for the Protection of Intellectual Property (the predecessor of the World Intellectual Property Organization), the EEC Commission, patent agents, lawyers, and industrialists.[109] Discussion continued over the next few years, and in 1973, sixteen countries met in Munich to sign the European Patent Convention. For the most part, the European Patent Convention (EPC) combined the rules of patentability set forth in the Strasbourg Convention with the "two-part" governance arrangement described in the EFTA Memorandum.

The EPC created a system that is largely in place today. It established the European Patent Office, a central bureaucracy similar to the PTO, to perform bureaucratic and judicial functions. The European Patent Office houses technical examiners who review patent applications according to the law and prior art, just as in the United States. These examiners have scientific or engineering training and are citizens of one of the EPO's member states. Once they are hired, they receive training in patent law through a dedicated patent academy and through an apprenticeship with a more experienced examiner. EPO examiners need to be competent in at least two of the organization's three official languages (French, German, and English) and be willing to learn the third. They also have to consider multiple categorical exclusions as well as the *ordre public* clause; however, these issues did not arise in a significant way before the biotechnology controversies began in the 1980s.

Like the PTO, the EPO also has internal appeals processes. Patent applicants unhappy with the EPO's decisions can appeal them first to the Technical Board of Appeal (made up primarily of technically trained personnel) and finally to the Enlarged Board of Appeal (made up primarily of legally trained personnel). Because the EPO is not part of the European Union, the EBA is the highest legal authority in the pan-European patent system. But once the EPO grants a patent and it becomes a "national" patent valid in one or multiple EPO member countries, it can be challenged through the national courts (and, if the country is in the European Union, up to the European Court of Justice).

An Administrative Council, made up of representatives (usually, from their patent offices) from all of the EPO's member countries oversees the EPO's activities. It plays a legislative role similar to Congress's involvement in the United States. But unlike the Congress, it is focused only on the EPO's activities and meets periodically at the EPO's main offices in Munich. Although it is not part of the European Union, the EPO often agrees to abide voluntarily by the European Parliament and Council's decisions in order to minimize variation between its work and those of national patent offices.

These aspects of the pan-European system emphasize the technical and legal dimensions of patents. But there is still room for consideration of moral and socioeconomic concerns. The EPC includes multiple exclusions as negotiated in the Strasbourg Convention, including the *ordre public* clause. It also has an opposition mechanism that allows any third party to challenge a patent within nine months of its issue. These challenges are limited to the grounds that the patent does not meet the criteria for patentability (which includes *ordre public*), that it is not sufficiently described, or that the granted patent goes beyond the initial application.[110]

These third parties can also appeal decisions to the TBA and EBA. Although the opposition mechanism was designed to improve the quality of patent decisions and reduce lengthy litigation after the grant of a patent,[111] it provided a potential space for the public in the patent system. After all, the EPC did not specify which third party could participate, and, as we will see, the EPO would not impose any restrictions when civil society groups began to file oppositions during the life-form patent controversies. This, coupled with the EPC's categorical exclusions, creates avenues for raising public concern, and it suggests that the European patent system's designers anticipated that the interests of the public would not always align with those of the inventor.

The United States has no real analog to the opposition mechanism. In 1980, it permitted third parties to initiate the "reexamination" of a patent, but it limits challenges to those who bring previously undiscovered "prior art" to the PTO's attention. In other words, it focuses on technical and procedural error, which means that challengers must have technical and legal expertise to participate. In 1999 and 2011, the United States created more opportunities for third-party intervention, but they are still limited to disputes over prior art.

The differences between the opposition and reexamination mechanisms

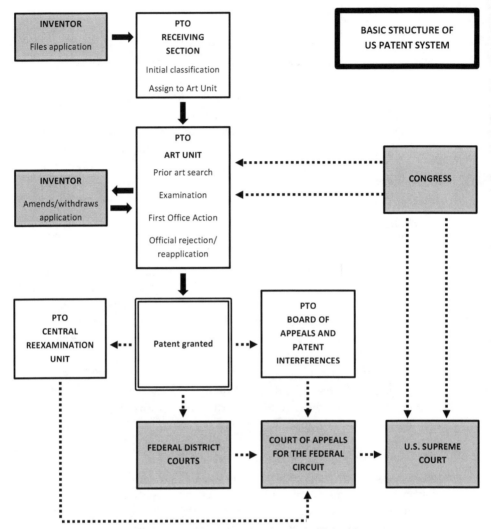

Figure 1.1. Process of patent examination and review in the United States.

reflect different market philosophies and different approaches to patents. Both jurisdictions use them to ensure that the "right" patents are granted. In the United States, this means that the PTO has done its due diligence and investigated all of the relevant technical evidence to ensure that the technology was indeed novel, inventive, useful, not obvious, and sufficiently described. But in Europe, ensuring that the "right" patent has been issued encompasses

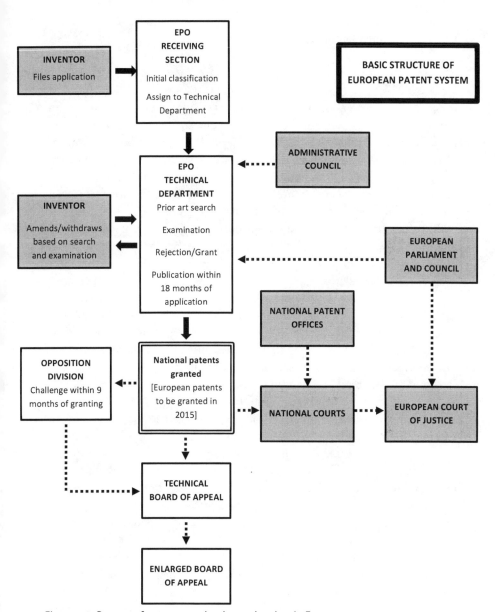

Figure 1.2. Process of patent examination and review in Europe.

moral and socioeconomic dimensions, and assumes the public beyond the inventor community has an important role to play.

CONCLUSION

In 1973, an advertisement for EPO examiners placed in the UK *Observer* began, "A sound patent system is the universally acknowledged means of encouraging the efforts of inventors and stimulating the development of new manufactures."[112] Indeed, by then, both the US and pan-European patent systems embraced this approach. Both seemed to treat patents as important innovation and market drivers. They also understood them as techno-legal objects, produced by bureaucracies driven by scientific, engineering, and legal expertise. For the most part, their rules and procedures were quite similar too, providing support for an increasingly global marketplace. While the EPO was a pan-governmental organization untethered to a formal government and with internal judicial and legislative processes, and the PTO was a technical bureaucracy within a federal executive with formal oversight from Congress and the courts, they seemed to understand patent governance in similar ways. Indeed, prospective inventors across the world could expect similar fates for their patent applications, whether they filed them in the United States or in Europe. And institutions like the World Intellectual Property Organization, established in 1967, and the Trilateral Co-operation, established in 1983 among the PTO, EPO, and Japan Patent Office to harmonize practices and discuss common concerns,[113] further assured innovators that they could indeed count on a global patent system.

But underneath this apparent harmony lay very different patent-system legacies and orientations. Europe had a long history of thinking about patents in socioeconomic terms, shaped by riots and disorder provoked by the early privilege systems and nineteenth-century controversies over the benefits of patents. In addition, it had developed multiple provisions based on the idea that patents could have mixed moral and socioeconomic impacts, and that the interests of patent holders and the public were not always the same. In these circumstances, the governments assumed and articulated a responsibility to step in. The US system, by contrast, was built with an over-arching emphasis on technical expertise and procedural objectivity. And while it experienced patent controversies in the early twentieth century, its decision makers and interests had developed strong enough positions by then that they were able to fend off alternative understandings of the patent system. These insiders distinguished between patents and patent-based monopolies, arguing that monopolies were created by patent holders and were therefore connected neither to the patents themselves nor to the government's responsibilities. Furthermore, they agreed that any government intervention would upset a patent system that generally worked. As a result, the US government's role was seen as quite limited: it made decisions to certify patentability based on science and law, but decisions regarding cost, access, and even the morality of commodification were left largely to the market or other policy domains.

Although international negotiations, including creation of the EPC, might have wiped away these differences, they remained. Indeed the European Patent Convention retained important markers of its legacy. It distinguished itself from US patent statutes and case law by maintaining some exclusions to patentability and articulating an *ordre public* clause. It also included an opposition mechanism that was conceived broadly enough to include, potentially, any citizen as a viable participant. These elements seemed minor at the time. The *ordre public* clause had engendered little discussion in the EPC negotiations, and it was unclear how violations would be assessed and by whom. The EPC had included fewer categorical exclusions than those that were written into national patent laws of the time. And the use of the opposition mechanism by civil society groups or other third parties seemed unlikely when there was little citizen mobilization related to innovation. But biotechnology and the controversies emerging around it would revive these provisions and force their reinterpretation in a new era. They would lead the United States and Europe to reassert and even strengthen their different understandings of patents, their governance, and the public interest.

2 CONFRONTING THE QUESTIONS OF LIFE-FORM PATENTABILITY

An entrepreneur excited about biotechnology's promise in the 1970s could reasonably expect that the United States and Europe would treat applications for life-form patents the same. The two jurisdictions now had very similar patent systems. They also seemed to have similar commitments to this area of research.[1] Scientists in both places were testing new techniques to cut DNA at specified places and recombine it (known as recombinant DNA, or rDNA), in order to analyze and manipulate living organisms at the molecular level, and they were developing means to map and sequence genes and identify their disease-causing mutations. They expected this work would not only enhance medical and agricultural research, but also produce immediate applications, including microbes that could be manipulated to clean up environmental pollution, engineered crops that could withstand drought, and genetic therapies for the most intractable human diseases.[2] There were, however, concerns brewing regarding biotechnology's risks.[3] In both the United States and Europe, governments, scientists, and citizens worried about the safety of humanity and the environment if these new life forms replicated uncontrollably and escaped the laboratory, and they were wary of the power that these new technologies implicitly gave to the scientific community in terms of manipulating life.

Meanwhile on both sides of the Atlantic Ocean, excited about biotechnology's promise, venture capitalists, universities, and industry sought to protect their investments by adopting a strategy that had been successful in other areas of technological development, including pharmaceuticals: they patented as much of their research output as possible. They were, after all, investing large amounts of money and other resources in the field, and these patents could help them establish market dominance for years to come.[4] They applied for US and European patents on the processes of making life forms (including rDNA) as well as on the equipment developed to do this research.[5] They also applied for patents covering the life forms themselves: scientists and their sponsors argued that molecular interventions had transformed these life forms into man-made inventions.

The United States and Europe would handle life-form patents very differently, and as they did so, they would embody different understandings of the patent system's role and of the public interest. They would also offer different definitions of life and its governance. In the United States, the question of life-form patentability was understood immediately as a legal one, focused on whether newly identified and manipulated organisms were discoveries of natural phenomena, which were not patentable, or inventions of new technology, which were patentable. US statutes allowed patents on "any new and useful process, machine, manufacture, or composition of matter,"[6] but case precedent prohibited patents on "products of nature."[7] So the central issue in the *Diamond v. Chakrabarty* case was whether living things should be treated as products of nature or as any other composition of matter. Here, the public interest was the development of a robust biotechnology industry and a predictable patent system. As the case proceeded, Jeremy Rifkin and other civil society challengers tried to shift the conversation to focus on the metaphysical aspects of allowing patents on *life*, rather than on products of *nature*. This might have required decision makers to consider the moral implications of such patents and cede decision making to the legislature. But these challengers were rebuffed rhetorically, told that they simply didn't understand how the patent system worked and that any interventions would interfere with the benefits of the free market. Eventually, the court implicitly determined that this genetically modified organism was not life at all.

In contrast, European decision makers interpreted the question as a political and policy one, to be negotiated by legislators and civil society groups. They focused on government's role in protecting the public interest, including morality, which could conflict with the inventors' interests. They worried about allowing patents on *life*, and particularly the implications for human and animal dignity. Citizens and policymakers used the *ordre public* clause as a tool to encourage discussion of the moral dimensions of patents, but despite what many legal scholars have argued, I demonstrate that it alone cannot explain either the debate or the final legislation. Legislators reinterpreted *ordre public* in terms of protecting human and animal dignity, but they also considered the impacts of patents on the livelihoods of farmers and scientists beyond their interpretation of the clause's scope. Ultimately, this produced legislation that articulated multiple exclusions to life-form patentability.

DEFINING PATENTS IN TERMS OF LAW
AND NATURE IN THE UNITED STATES

The life-form patentability question arose first in the United States, with Ananda Chakrabarty submitting a patent application on his genetically engineered microorganism in 1972. After reviewing the application, the PTO decided that the microorganism was unpatentable because it was a "product of nature."[8] Chakrabarty and General Electric appealed this decision, officially made by PTO Commissioner Sidney A. Diamond, to the Court of Customs and Patent Appeals, the federal court that heard all appeals of patent cases (which eventually became the Court of Appeals for the Federal Circuit, or CAFC). The CCPA overturned the PTO's decision and allowed Chakrabarty's patent in 1978. As it did so, it ruled that there were no legal barriers to the patentability of living organisms. The case had proceeded according to the customary process, and the PTO was content to use the appeals court's opinion to begin issuing life-form patents.[9] The Justice Department (DOJ), however, wanted clearer guidance and asked the Supreme Court (SCOTUS) to review the matter. SCOTUS took the case, and when it did, life-form patentability shifted from a bureaucratic question to a legal issue of public importance. As SCOTUS adjudicated whether life forms were patentable, it was now focused on the issue of whether they should be understood in terms of nature or in terms of technology.[10]

The DOJ's desire for clear guidance made sense in the context of the emerging biotechnology industry. The patentability of life forms would influence business strategies and likely the shape of the emerging industry overall. It would even affect the universities whose employees were doing a lot of biotechnology research.[11] It is therefore not surprising, given its focus and the relative importance of the courts in shaping the patent system, that the DOJ sought additional guidance from the Supreme Court, but it is notable that there seemed to be no controversy regarding whether the matter should be restricted to the legal arena. As discussed in chapter 1, Congress had discussed patent-related matters before, and it certainly had the jurisdiction to do so. It had restricted patents related to atomic weapons in the 1940s and 1950s, and created a stand-alone (sui generis) system for governing plant patents as well.[12] But these interventions were rare. And Congress did not get involved this time, underlining the US approach to patents in legal, as opposed to policy, terms.

Stimulating Innovation and the Market

The Supreme Court heard the *Diamond v. Chakrabarty* case in March 1980. For the most part, it proceeded as a normal case focused on patentability. Familiar interests in the patent system, who had given testimony in previous patent-related Congressional hearings,[13] submitted amicus briefs. Briefs came from patent-law associations, the US biotechnology company Genentech, universities (including the University of California), scientific associations, the US Pharmaceutical Manufacturers Association, and a molecular biologist.[14] Amicus briefs articulate the positions of parties interested in a particular court case, offer additional information and arguments, and provide a sense of public opinion to the court.[15] Anyone can submit one, but those with access to legal expertise, who can craft the documents to fit the correct format and parameters, usually submit them. In *Chakrabarty*, all but one of the fifteen briefs supported patentability.

By and large, these pro-patent briefs defined Chakrabarty's organism as a technology. He had clearly manufactured the organism, they argued, by splicing in new DNA, and therefore it was patentable and not a product of nature. "It never existed until the respondent made it," argued the American Patent Law Association.[16] They also observed that Congress had allowed patents on natural products for many years. Finally, they noted that the requirements for patentability enshrined in statutes and case law made no distinction between living and nonliving organisms.[17] In fact, on multiple occasions, Congress had considered the scope of patentability and had never explicitly excluded living organisms from consideration.[18]

Supporters of Chakrabarty's application also framed their arguments in terms of economic growth and scientific and technological progress. The American Patent Law Association, for example, stated, "The public, as a result of the [biotechnology] industry, now can enjoy the fruits of that labor but the public had nothing to enjoy, freely or otherwise, prior thereto."[19] The message was clear: prohibiting life-form patents would contravene the patent system's purpose of stimulating innovation, which could have enormous negative impacts.

Further, pro-patent amici argued that if the US banned patents on living organisms, the costs to America's global competitiveness would be extremely high. Citing newspaper and magazine reports, political speeches, government reports, and academic articles, they suggested that patents on living organisms were necessary for the growth of the biotechnology industry and, ultimately, for the growth of the nation. Genentech noted that "the fear has been widely expressed that the United States increasingly is losing its technological lead, and that the loss of that lead can be expected to severely impact America's balance of payments and other indicia of economic health. . . . The encouragement of domestic innovation is important, and that can best be done by a strengthened patent system."[20] The Pharmaceutical Manufacturers Association was more explicit: "European governments are actively promoting genetic engineering research and some US investment and research activity is moving overseas. The potential exists, therefore, for the US to lose its hold on this technology, and this potential should not be lightly dismissed in considering the practical impact of the issue presented by this case."[21] There is some question about whether the Europeans were indeed promoting genetic engineering actively at the time, and this area of technology would soon become more controversial in Europe than it was in the United States.[22] But this rhetoric was powerful, not only because it articulated the market-making ideology dominant in the United States, but also because the country had a long history of understanding patents as important drivers of innovation and markets.

Imagining Patents Differently

The People's Business Commission (PBC, led by Ted Howard and Jeremy Rifkin) and its allies filed the lone opposing brief. This brief treated patents as having enormous moral and ecological impacts across generations. With this came very different ideas of the appropriate decision makers and of the public's role: the brief argued that patenting life was a matter of policy that Congress, rather than the Supreme Court, should consider.

The PBC's concerns evolved from deep skepticism of both technology and business. Today well known as a biotechnology critic,[23] Rifkin began his career as an antiwar activist but used the occasion of America's bicentennial celebrations to create the People's Bicentennial Commission (renamed as the People's Business Commission in 1976). A small organization run on a shoestring budget and with the free labor of college students, the PBC saw itself as connected to the civil rights and antiwar movements of the 1960s and 1970s in spirit and in its desire for fundamental social change, but it had a different goal: fundamental economic change that would bring economic and political power back into the hands of average citizens.[24] In order to accomplish this, it used theatrical tactics familiar to its sister social movements of the time. In 1976, for example, it sent letters to the wives and secretaries of the CEOs of major corporations, offering a $25,000 reward to anyone who could provide "concrete information that leads directly to the arrest, prosecution, conviction, and imprisonment of a chief officer . . . for criminal activity relating to corporate operations."[25] However, it also sought supporters beyond "traditional radical constituencies" and therefore produced educational reports and books that could be distributed widely (and used well-known publishers including Bantam, Simon & Schuster, and McGraw-Hill).[26]

Rifkin had learned about the potential dangers of biotechnology research from friends associated with the left-wing, antiwar group Science for the People (SftP). SftP was made up of graduate students and professors who advocated the development of science according to social, rather than military or industrial, priorities.[27] It included eminent scientists such as Richard Lewontin, who pioneered the development of gel electrophoresis, which was a pivotal step in the development of biotechnology research. It had become concerned about the risks of rDNA both for laboratory workers and for the environment, and was unhappy with the lack of public participation in discussions about biotechnology's governance. One of a few scientist-led protest movements that operated during this period, StfP helped the PBC develop its position and strategy regarding biotechnology.[28] The two groups also worked together on a few occasions, including a joint protest at a meeting of the National Academy of Sciences convened to discuss rDNA research.[29]

Soon, biotechnology research became the PBC's main focus. In 1977, Howard and Rifkin published *Who Should Play God? The Artificial Creation of Life and What It Means for the Future of the Human Race*.[30] The book argued that the new field of biotechnology raised moral questions about the meaning of life that were going unanswered, that the new field could cause serious damage to health and to the environment, and that the emerging governance of biotechnology did not reflect democratic principles. It also suggested that the coming biological revolution had the power to change the world more profoundly than even nuclear weaponry, which at the time was a major area of concern for many on the political left and for many scientists.[31]

Who Should Play God? was also one of the first publications to call attention to the potential dangers of corporate control over biotechnology. The PBC's very mission rejected the idea that a top-down approach to economic power would help average citizens. It argued that a patent placed market control in the hands of a single owner, giving her complete control over whether and how to use the technology. This included control over who else could use the technology and how they could do so, which could shape future research and technological development. Rifkin and Howard were particularly worried about biotechnology patents, because they would allow a small handful of companies to be privy "to the secret of life and how to manipulate and change it."[32] They argued that the pharmaceutical and nascent biotechnology industries would, if the new field was successful, be able to essentially control biology through patents. The book observed, "Not a single member of [the NIH rDNA Advisory Committee (RAC), which reviewed all rDNA experiments to ensure that adequate safety precautions were taken,] questioned the right of private corporations to assume a proprietary ownership over any 'new life forms' that they commercially develop. This unanimous agreement over the patent issue was accepted virtually without reservation in House and Senate hearings on industry involvement in genetic engineering in the late spring of 1977, thus establishing a precedent which could have serious implications five or ten years from now."[33]

These concerns drove Rifkin and Howard's intervention in the *Chakrabarty* case. The best way to have their voices heard in the case was likely through an amicus brief, even though they had no experience with the patent system previously.[34] But the PBC did have in-house lawyers and a history of unconventional tactics, particularly in creating political opportunities when there seemed to be none. In the years after *Chakrabarty*, for example, it would file suit against the US Department of Agriculture (USDA), arguing that the USDA had not conducted adequate review of the potential environmental impacts of genetic engineering research on animals before funding it. [35] It also tried to force the NIH's rDNA Advisory Committee to consider the broad implications of biotechnology research by turning its review process into an opportunity for activism. The PBC submitted a proposal for interspecies germ-line genetic transfer; its proposed work would mix DNA from different species and ensure that the resulting genetic changes would be passed down to future generations. The new organisms, combining DNA from different types of animals, or between animals and humans, could be scientifically and medically useful. But, the PBC hoped that the idea of the research would be so shocking that it would generate public scrutiny of the RAC's activities. However, policymakers had structured the RAC to avoid the kind of deliberation that the PBC wanted, and it received hundreds of letters opposing (with some even ridiculing) the PBC's attempt.[36]

The PBC's brief in *Chakrabarty* argued that policymakers should evaluate the patentability of life forms in terms of their potentially broad social and environmental implications. With this reframing came different understandings of patents and their governance, including the relative responsibilities of the public, experts, and the government. It suggested that patents had environmental and social effects, and that those who set patent policy were responsible for them. To make this argument, it used the historical analog of special "plant" patents that the United States had allowed since 1930. Plant patents conferred a more limited set of intellectual property rights to the inventor than the utility patents that are the most common form—the ones that were at issue in the *Chakrabarty* case and that are the subject of this book.[37] The PBC argued that even these plant patents, which allowed farmers to save their seed across generations, had led to the corporate consolidation of the agricultural industry. This consolidation had, in turn, eliminated plant varieties, created monocultures that could be completely destroyed by one pest or disease, and disadvantaged small farmers.[38] It supported this argument with reports from development organizations and from a National Academy of Sciences committee focused on the "genetic vulnerability of major crops," which demonstrated that plant patents had resulted in fewer companies in the agricultural sector. The brief concluded: "Because plants are the only living organisms now patentable, the above mentioned consequences of plant patenting must be seriously considered as a possible pattern that may be repeated should other forms of life be deemed patentable by this Court."[39] This argument was notable because it introduced a different understanding of patents and because it suggested that historical comparison as well as health, environmental, and socioeconomic research done on plant patents were relevant to the debate.

The PBC brief also suggested that patents had moral consequences that the government was obligated to consider. Patents on living organisms would change the meaning of life and the relationship between humans and nature in profound ways, it stated: "All living material will be reduced to an arrangement of chemicals, or mere 'compositions of matter'" for "our manipulation, exploitation, and transformation."[40] Although the case focused on a microorganism, it noted further, it would be impossible to distinguish between lower- and higher-order organisms for the purpose of patentability; thus, all beings could potentially be owned. It would strengthen dominion over life forms through property and ownership.

To make these arguments, it cited the work of scholars who were not traditional participants in the patent arena, including French philosopher Jacques Ellul, who had written critically on the relationship between humans and technology; bioethicist Leon Kass, who years later would lead the US president's Council of Bioethics under President George W. Bush; and respected microbiologist Salvador Luria, who was a left-of-center political advocate. These scholars warned against turning life forms into "industrial products," arguing that this would "violate human dignity."[41] Kass noted, for example, that "'increasing control over the product is purchased by the increasing depersonalization of the process.'"[42] Treating life forms as technologies ready for the marketplace, he argued, would devalue life overall. Here too, just as with its invocation of the history of plant patents, the PBC implicitly claimed that the expertise of these unfamiliar scholars was relevant to the patent system.

On the basis of these references, the PBC and its allies argued that the life-form patent question could not be simply answered with technical knowledge and legal interpretation of the "product of nature" doctrine, as if it were similar to the dozens of patentability questions that the Supreme Court had answered in the past. But they also saw a central role for the public, noting in its brief, "The question of whether the public will be well-served by the patenting of living organisms and the technology of genetic engineering should most properly be left to the public-at-large and its elected representatives."[43] These new claimants challenged the prevailing market-making approach by treating patents as matters of social value that could and should be separated from economic value. As they did this, they disputed the idea that an unfettered market and a patent system that simply facilitated it would necessarily produce social good. They also argued that these matters had to be addressed and balanced by a broad set of experts and citizens rather than by a narrow group of experts.

A brief from such unconventional parties might make it hard for its arguments to gain attention from the justices and their clerks, who were accustomed to players who were patent system insiders and to certain kinds of conventions and arguments. Both the courts and other amici could simply have ignored Rifkin's brief. But they did not, perhaps because Rifkin had managed to stimulate some media attention,[44] and because biotechnology was becoming increasingly controversial: the US Congress had held multiple hearings on the regulation of rDNA research, and municipal governments (e.g., those in Cambridge, Massachusetts, and Ann Arbor, Michigan) even banned rDNA research for a short period.[45]

Reasserting The Techno-Legal Understanding of Patents

The pro-patent amici dismissed the PBC's approach summarily. As they did so, they not only reasserted the customary ways of thinking about patents in the United States but also dismissed the idea that government and the public had active roles to play. They also reiterated the idea that life forms were similar to any other patentable technology. The American Patent Law Association, for example, argued that "the Patent and Trademark Office well knows that its function is to *examine* inventions presented to it for compliance with the patent statutes, not to *regulate* hazardous research [emphasis in original]."[46] Of course, a domain that only certified patentability had limited scope for intervention. These amici also tried to maintain a narrow idea of what constituted relevant knowledge and expertise in the patent system. The Pharmaceutical Manufacturers Association, for example, distinguished between the history of "plant" patents and the "utility" patents that were at issue in the *Chakrabarty* case. Because they were debating different kinds of patents, "the practical success or failure of the Patent Acts covering plants is irrelevant to the issues in this case."[47] As they argued that historical analogs were irrelevant, the association rejected both the evidence and the idea that the PBC understood how the system worked.

The Supreme Court agreed with the pro-patent amici. In a 5–4 decision, it decided that Chakrabarty's organism was a patentable technology, interpreting congressional intent broadly to allow patents on "anything under the sun made by man."[48] As it made this decision, the majority reinforced the idea that patents only influenced society by driving innovation and therefore markets. Chief Justice Warren Burger, who wrote the opinion, stated, "Whether respondent's claims are patentable may determine whether research efforts are accelerated by the hope of reward or slowed by want of incentives, but that is all."[49] He specifically rejected the idea that patents had other impacts on risk or on the direction of innovation, noting, "The grant or denial of patents on micro-organisms is not likely to put an end to genetic research or to its attendant risks. The large amount of research that has already occurred when no researcher had sure knowledge that patent protection would be available suggests that legislative or judicial fiat as to patentability will not deter the scientific mind from probing into the unknown any more than Canute could command the tides."[50] As he characterized patents as focused only on the pace of innovation, he also defined appropriate governance as narrow, focused simply on "the introduction of new products and processes of manufacture into the economy."[51]

This outcome becomes more meaningful when we step back and consider that Congress could also have intervened afterward, but did not. Indeed, the majority opinion invited it to do so and the minority opinion, written by Justice William Brennan, argued that the Supreme Court simply could not authorize the patentability of life forms. Rather, Justice Brennan stated, Congress's explicit allowance of plant patents in 1930 suggested that it would expect to have this power over utility patents on life forms as well, especially because "the composition sought to be patented uniquely implicates matters of public concern."[52] But Congress seemed content to treat the question of patentability as a legal and technical matter that was now settled.

Neither the House nor the Senate responded with official hearings, and no congressperson proposed legislation. The case had gone through the proper channels and was resolved. To reopen the issues would have violated the patent system's predictability and procedural objectivity, which could diminish trust in a system that the public largely believed in. Indeed, Congress and the scientific community seemed enthusiastic about the benefits of life-form patents for the US biotechnology industry and the economy as a whole. In August 1980, the House Committee on Science and Technology, the American Society for Microbiology, and the Congressional Clearinghouse on the Future sponsored a public forum on the subject, but it did not explore the issues that the PBC had raised. [53] Rather, it focused on the scientific and industrial potential of expanding patentability to involve life forms. The episode essentially ended there, and the PTO soon began to allow these patents.

Why were US decision makers so resistant to thinking about patents in moral and policy terms? They had, after all, framed previous discussions about blood and organ donation in terms of the ethics and social implications of commodification.[54] I think the answer lies in the focus of the debate within the US *patent* system, which had a long history of considering matters in legal and scientific terms: if the system made decisions objectively, then the public would benefit. It had also traditionally aligned the interests of inventors and the public. In addition, bureaucrats, legislators, and organized interests had worked hard to reject the idea that patents had downstream implications. As we saw in chapter 1, the consequences of monopolies were not seen as the result of patents or the responsibilities of the patent system; rather, they were the responsibilities of patent holders. As a result, decision-making and organized interests could argue that any moral implications of commodification were unrelated to patents. In fact, this is perhaps one of the reasons that even the moral utility doctrine had very minor impact. By contrast, discussions of the donation of biological things have been linked to medicine and public health, which had by then become accustomed to considering moral issues and potential conflicts between health and commerce.

In fact, Congress reinforced its vision of patents as pivotal innovation and market drivers later in 1980 when it passed the Bayh-Dole Act. Bayh-Dole allowed universities, small businesses, and nonprofit organizations to patent technologies that resulted from government-sponsored research.[55] It was designed to resolve confusion over the ownership of government-sponsored inventions, which industries, universities, and the government worried was stifling innovation. Devolving ownership of potential patents to market players, including universities, the logic went, would inspire more companies to purchase licenses to these patents and develop the related technologies for the market.[56] It was a boon for universities, which housed an enormous amount of such research. They could now unambiguously patent the fruits of this work, make licensing agreements as appropriate, and reap revenues from commercializing patented inventions.[57] Biotechnology companies also stood to benefit. They could license university inventions exclusively, without fear that competitors might be building on the same work.

DEFINING PATENTS IN TERMS OF LIFE AND POLICY IN EUROPE

Initially, Europe's reaction to life-form patents was more permissive than that of the United States. When the EPO and national patent offices in Europe received applications covering life forms in the 1970s, they treated them like any other.[58] Examiners applied the rules set in national patent laws and in the EPC, and began to grant them.[59] While these laws prohibited patents on "discoveries," microbiological processes, and plant and animal varieties, European patent offices did not seem to see them as a barrier to life-form patents.[60]

But the growing European biotechnology industry still wanted explicit authorization. They worried, particularly in the wake of the *Chakrabarty* case in the United States, that the slight differences across European patent laws would lead to different decisions.[61] This would hurt the fledgling biotechnology industry by making it more difficult and more costly to develop pan-European business and innovation strategies, which would in turn hurt the emerging European common market. Inventors might choose to do their work and sell their products in countries that had stronger patent protections.[62] It was particularly important to maintain the European biotechnology industry's competitiveness, they argued, since the United States had been approving biotechnology patents for years.[63] The biotechnology industry could have asked the patent offices to harmonize their examination approaches through transnational agreements, but instead it took a more direct and formal route that would have more authority. It convinced the European Commission, the executive arm of the European Union, to develop legislation on the issue. So, in 1988, the Commission (specifically, the directorate general responsible for industry) proposed the EU Directive for the Legal Protection of Biotechnological Inventions, known as the Biotech Patent Directive (BPD). It was designed to harmonize European patent law in the area of biotechnology, and included language that would extend patentability to clearly include life forms. Once the European Parliament and Council passed it, the BPD would be valid in all EU member states.

The EPO agreed to abide by it voluntarily. Although it is difficult to know for sure why its officials decided to adopt the BPD even though the EPO is not part of the European Union apparatus, it was likely in its best interest to align its policy to those of national patent offices in order to facilitate a European common market. It would also ensure the bureaucracy's political legitimacy. It was, after all, a pan-governmental organization only indirectly tied to a voting public. Although this may not have mattered much at the time, it would become quite important over the next few years, as the EPO became a target for civil society groups opposed to its patent decisions. It would be able to refer to this EU legislation for additional support and guidance.

When the European Commission submitted the BPD to the legislature for deliberation and a vote among elected representatives of European citizens, it clearly saw the process as perfunctory. Its proposed BPD merely focused on the uncontroversial matter of patents, which stimulated the European economy. By taking this approach, the Commission echoed the PTO and organized interests in the United States. In its introductory text, it agreed with the biotechnology industry that the proposed law was an important step toward completing the European common market: "This proposal is one of the measures aimed at providing industry with the ability to treat the common market as a single environment for their economic activities and to create the conditions necessary for the proper functioning of the common market."[64] In the absence of such a law, the Commission argued, neither the common market nor EEC industries would function properly. It also observed that European law was out of date, given the rise of biotechnology.[65] For example, the distinctions made in the EPC — among them the patentability of "microbiological processes and their products," the unpatentability of "plant and animal varieties," and "essentially biological processes for the production of plants and animals" — no longer made sense in the context of modern biotechnology: biotechnology was changing society's understanding of these categories, including what was "essentially biological."

The proposed BPD also stated that the law would strengthen EU initiatives to foster research and development in the area of biotechnology. Researchers involved in collaborative projects across the EU would not have to worry about "appropriate industrial protection."[66] The Parliament's apparently minimal role likely encouraged the Commission's legislative strategy. In 1988, it was a consultative body. Only the Council, made up of leaders from the EU's member states, needed to approve the BPD. In sum, the Commission saw life-form patents just as the US Supreme Court and Congress had, as techno-legal objects that could stimulate markets and innovation, and likely expected quick and uncontroversial passage of the BPD.

In a surprise, the European Parliament challenged the European Commission's approach almost immediately. It asked its Economic and Social Committee (ESC), a standing assembly made up of representatives from employers' organizations, trade unions, and other interest groups, to review the draft BPD. Although this kind of consultation was becoming customary in Parliamentary matters,[67] and was common in many of the corporatist states of Europe, such groups did not often have the opportunity to weigh in on patent-related policies. They had not weighed in on the crafting of the European Patent Convention, for example. But for parliamentarians, the members of the ESC could bring important knowledge to bear on the patentability of biotechnology and could lend legitimacy to their decision.

In April 1989, in the first sign that the European debate would look very different from the US one, the ESC issued its opinion. It criticized the Commission for considering only the interests of industry and called for the incorporation of perspectives from a wider variety of stakeholders (including, for example, farmers, breeders, consumers, and researchers).[68] If it had considered these perspectives, the ESC argued, the Commission would realize that the proposed BPD had important ethical implications that needed to be taken into account, and that it would have detrimental effects on agriculture, science, and the economy. The ESC, unlike the Commission, saw life-form patents as tied to a variety of implications and therefore requiring robust discussion among a much broader array of interests.

In response, the Parliamentary Committee on Legal Affairs and Citizens' Rights (CLACR), which was responsible for the initial review of the BPD, designed a comprehensive assessment of the proposed law. Vice Chair Willi Rothley, a German member of the European Parliament (MEP) since 1984 from the European Socialist Party, served as rapporteur. He asked three other Parliamentary committees—the Committee on Economic and Monetary Affairs and Industrial Policy; the Committee on Energy, Research, and Technology; and the Committee on Agriculture, Fisheries, and Rural Development—to issue reports on the matter to supplement CLACR's analysis. (Eventually, he would ask the Committee on Development and the Committee on Environment, Public Health, and Consumer Protection for their opinions as well.)

Rothley's approach to analyzing the BPD immediately challenged the narrow and technical definition of patents put forth by the Commission. Inviting opinion from a diverse set of committees, he opened up the debate—and consideration of the implications of patents—in a way that the Commission had not foreseen and that US politicians had not done. This kind of engagement was new for the European Parliament, which began in the 1950s simply to represent public opinion with members drawn from national parliaments. Europeans only began to elect members directly in 1979, and the Parliament was gaining power and responsibility slowly.[69] In many respects, just as the EPO was becoming an important space for the construction of a unified Europe, biotechnology was proving to be one of the first testing grounds for the European Parliament and for regulatory policymaking at the European level. Earlier in 1989, CLACR had guided the passage of a Parliamentary resolution encouraging the establishment of a special commission to review the impact of genetic engineering. In that same year, the Parliament would also approve the EU's first legislation regarding the "deliberate release" of genetically modified organisms.[70]

Rothley reinforced this broad understanding of the implications of patents by convening a public hearing in May 1990 that included testimony from eight witnesses, including agricultural and industrial representatives, patent lawyers, and the leader of Genetic Resources Action International Network (GRAIN), an advocacy group concerned about the loss of genetic diversity worldwide.[71] Rothley's invitation to GRAIN was notable, because the organization was becoming a vocal critic of life-form patents and because it focused on environmental and socioeconomic issues, including the impacts of agricultural biotechnology on farmers and biodiversity.[72] At the hearing, GRAIN's coordinator, Henk Hobbelink, presented concerns about the effects of patents on biodiversity, food security, corporate concentration, and the meaning of life. By inviting Hobbelink's testimony, Rothley seemed to see these issues as relevant to patents, and to understand decision making as requiring expertise from civil society actors who were new to the patent system.

Over the next few months, each Parliamentary committee submitted its assessment of the BPD. All of them raised concerns that echoed the PBC's arguments in *Chakrabarty*, but the Agriculture and Development Committees were the most opposed. Their concerns fit into five categories. First, many committees worried that the socioeconomic costs borne by farmers, breeders, and consumers would outweigh the benefits of economic growth at a broad scale. The Agriculture Committee argued that if life forms were patentable, "Independent breeders will no longer be able to use varieties and races freely for further innovation as their access to patented genetic information and techniques will be subject to delays (three to four years after the grant of the patent, according to the draft BPD) and payment of royalty fees for the whole patent period (20 years). The economic strain on the breeding sector will be such that most independent breeders will simply go out of business, if they are not bought up by multinational firms first."[73]

The Commission had focused on the jobs and profits that the BPD would bring through the development and commercialization of new biotechnologies. This emphasized the law's market-making potential. By contrast, these committees were pointing out potentially uneven distributional impacts by suggesting that farmers would lose financially; bringing plants into the utility patent system would disrupt the plant-variety protection system already in place (the analog to the US plant-patent system), which allowed farmers to save seed year after year. Committees also argued that European farms were comparatively small and would likely shut down if they had to bear the costs of buying patented seeds each year.[74]

Second, some committees echoed the PBC's concern that a strengthened patent regime would lead to consolidation of agriculture worldwide, which would create environmental, social, and public-health problems. Allowing patents on seeds, for example, would provide a competitive advantage to patent holders, which would lead to fewer plant varieties. Ultimately, this would hurt global biodiversity. This could become more problematic if one pest wiped out a variety, leaving large regions without access to a particular crop. The Development Committee noted that "wrong decisions in agriculture which restrict the diversity of species can have catastrophic effects on the security of food supplies, as has already become apparent with the displacement of traditional crops (millet) by western field crops (maize, wheat) in some African countries. ([Potential impacts include] lower resistance to disease in plants and animals; emphasis on high yielding varieties and races without reference to regional and ecological sustainability)."[75] And by restricting "the access of these countries to their own flora and fauna,"[76] patented seeds would also make farmers more dependent on corporations in the industrialized world.

Third, three of the committees worried about allowing patents on animals. Some feared that such patents would encourage the production of animals that would experience considerable suffering. MEP Rothley himself, for example, cited the development of the "Beltsville Pig," which was genetically engineered to produce extra growth hormone.[77] This pig suffered from arthritis and was more susceptible to infection. Others predicted, as had the PBC in the United States, that allowing animal patents would alter the relationship between humans and animals fundamentally.[78] It would create a sense of ownership and dominion like never before.

Fourth, all of the committees agreed that the BPD needed stronger language to ensure the protection of human dignity. For some, this simply meant that it should explicitly exclude patents on "human beings."[79] But for most, this meant categorical exclusions on isolated or manipulated parts of human beings, including human genes, cell lines, organs, and tissues.[80] The idea of a right to human dignity had deep roots in Europe.[81] In the wake of its revolution, France had extended this right, previously enjoyed only by the elite, to all of its citizens. In nineteenth-century Europe, it was used to abolish slavery and to motivate labor and other social movements. It was enshrined in Germany's 1949 Basic Law. Governments were responsible for upholding this right, which sometimes required them to limit the free market. We saw this in the early development of the patent system, and it had now reemerged with the consideration of life-form patents.

Finally, unlike the PBC, committees raised concerns that the BPD would actually stifle innovation in Europe by placing "severe restrictions on the free exchange of scientific resources and information."[82] Europe, unlike the United States, already had an experimental use exemption that allowed researchers to use patented inventions without a license.[83] But Parliamentary committees worried that the new legislation would weaken it. This would directly affect not only public research institutes and universities, but also farmers and breeders who developed improved agriculture through generations of cultivation, and citizens who might benefit from cheaper and better medical and agricultural biotechnologies. The Agriculture Committee report noted, "Patent lawyers are already dictating research priorities in the US biotech industry in open conflict with senior scientists, as monopoly interests take precedence over medical, agricultural, or environmental logic."[84] European universities did not have the same growing culture of commercialization as their US counterparts, and most European countries did not have any analog to the Bayh-Dole Act.[85] It must have seemed particularly strange, then, to allow patents on biotechnology processes and products developed early in the research process and often at universities.

Taken together, these committees had an expansive understanding of the implications of patents and of the role of government in shaping them. They saw patents as assigning control to a single owner, who could shape farming, the environment, research, and even public morality. This was, of course, a very different understanding from the one articulated by the European Commission.

Defining the Implications of Patents and the Ordre Public *Clause*

CLACR Vice-Chair Rothley reacted to many of the committees' objections by suggesting that these were simply characteristics of the modern condition. He accepted the idea that patents were connected to problems of agricultural industry consolidation and global biodiversity, but he suggested that changes to patent law alone could not solve them.[86] Therefore, they were not worth addressing in the BPD. In so doing, he tried to draw a distinction between those *direct* issues that could only be addressed in the patent system and those that were *indirect* and could be addressed through other policy domains.

Rothley did take seriously what he called "ethical" concerns regarding patents related to animals and humans, presumably because he saw those impacts as direct. Invoking the *ordre public* clause, which had fallen into disuse and had previously been interpreted—as I discussed in chapter 1—in terms of socioeconomic and national security harms, he suggested that the BPD should include explicit prohibitions on the following types of patents: (1) those that covered human beings; (2) those that failed a "comparative assessment" of usefulness versus risk; (3) those that involved "unnecessary" animal suffering; and (4) those that covered chimeras (organisms that combined genetically different cells from two animals).[87] He reasoned that such prohibitions would require examiners to consider the utility of life forms more seriously and discourage innovation in certain areas and therefore the creation of immoral markets. As he brought back the *ordre public* clause, he defined it in a new way, focusing on human and animal dignity and referring to it in terms of ethics: "Those responsible for [granting patent rights] must . . . be reminded of their ethical responsibilities. The dignity of human beings, their genetic identity, and the protection of animals against unnecessary distress must continue to be assured."[88] Rothley argued that patents had direct moral impacts by commodifying life forms, creating unethical markets, and infringing upon human and animal dignity. These needed to be distinguished from indirect environmental and socioeconomic implications, which he suggested had multiple causes. So while decision makers in the United States had focused on the patent system's stimulation of innovation, Rothley articulated a broader scope that included attention to human dignity.

Rothley's interpretation of the *ordre public* clause in terms of protecting societies from immoral inventions and markets, with morality defined in terms of human and animal dignity and welfare, reflected both the rise of bioethics and ongoing European controversies over human and animal research. Although bioethics had been more quickly institutionalized in the United States with the creation of academic programs and departments, formal policies, and the establishment of specialized advisory committees,[89] it was also establishing itself in Europe as a field that claimed expertise in dealing with public concerns about new science and technology.[90] In November 1991, for example, the European Commission created the Group of Advisors to the European Commission on the Ethical Implications of Biotechnology (GAEIB, the precursor to the European Group on Ethics in New Science and Technology, whose expertise would be codified later in the final text of the Biotech Patent Directive). Members included law professors, biologists, a physician, a philosophy professor, and a theologian, and they provided official written opinions on ethical matters in biotechnology when asked to do so by the Commission.[91]

While the United States had established similar expert advisory committees to contend with the ethical concerns raised by biotechnology, European governments had been more proactive in structuring regulations with these issues in mind. For example, at the time of the BPD discussion, many European governments were grappling with the ethical dimensions of embryo research, which began in the early 1990s after the success of in vitro fertilization. Embryos had stem cells that could, in theory, regenerate into any organ or tissue in the body, and could therefore be immensely helpful for medicine. But many citizens deemed this area of research ethically problematic, because it required the destruction of human and animal embryos, which were understood as higher life forms than other cells. While the United States had responded tepidly, with reports from expert advisory committees and some temporary restrictions over federal funding in this area,[92] European governments had responded more forcefully and systematically. The UK set up the Human Fertilization and Embryology Authority to review and regulate research in this area, while other countries passed laws that restricted research significantly.[93] Most strict was Germany's 1990 Embryo Protection Law, which banned all embryo research.[94]

We can see a similar pattern in how the US and Europe handled animal welfare concerns in research.[95] The United Kingdom's Animals Act of 1986, which covered the use of animals in laboratory research, has been referred to as the "tightest system of regulation in the world."[96] Switzerland passed a national referendum in 1992 that recognized animals as "beings,"[97] and eventually the European Union would pass multiple directives regulating the use of laboratory animals in research.[98] The United States, meanwhile, established guidelines for federally funded animal research but tended to not interfere with industrial activity. In addition, the US laws were generally weaker.[99] In sum, both European bioethics discussions and European laws related to human and animal research had engaged in proactive market-shaping activities. By focusing on whether and how life-form patents might create morally and socially problematic markets, the BPD debate was similar.

Ultimately, Rothley's response defined patents in both bioethical and socioeconomic terms, and assumed that the European Parliament, patent bureaucracies, and the courts had the responsibility to step into the patent system to shape these impacts. Furthermore, while Rothley had been reluctant initially to deal with the socioeconomic implications of monopolies, he proposed amendments that would address some of them. In addition to fleshing out the *ordre public* clause in bioethical terms, he suggested weakening patents on plants. He justified this by arguing that the framers of the EPC were clearly concerned about the socioeconomic effects of intellectual property for agriculture. He included a strong farmer's privilege, which would allow farmers to continue to reuse seed over generations without fear of patent infringement, just as the plant-variety protection system allowed, and he proposed an explicit exemption for researchers to use patented materials. He sought to allow farmers and scientists to innovate outside the market, and wanted to ensure that their livelihoods would not be drastically affected. In addition, by limiting plant patents, he hoped to ensure wide access to food, a value that many of the pre-EPC European patent systems had held.

The Full Parliament Weighs In

When the full European Parliament took up the Biotech Patent Directive in April 1992, it inspired vigorous debate. Most parliamentarians from across the political spectrum seemed to agree that patents influenced public morality and envisioned a governance system that took responsibility for these effects. But while parliamentarians agreed that the European Commission had paid inadequate attention to the moral significance of the BPD, there was disagreement over how they should define morality and address it through policy. MEP Margarida Salema, a Portuguese member of the conservative Liberal Democrat and Reform Party, adopted an approach similar to Rothley's: "It is important to combine these problems associated with the traditional concepts of patent law with the ethical aspects arising from values such as human life and dignity or the well-being of animals."[100] MEP Hiltrud Breyer of Germany argued that the moral implications of commodifying life required widespread exclusions to patents on life forms beyond animals and humans.[101] Breyer represented the Green Party, which had emerged in 1970s West Germany to focus attention on environmental issues, including the harms of nuclear power.[102] By the 1980s it had begun to advocate strict controls on genetic engineering research and development, and it would take a leadership role in the activism against life-form patents.

Others, like MEP Alain Pompidou, a French member of the conservative European Democratic Alliance and professor of histology, embryology, and cytogenetics at the University of Paris, suggested the involvement of "a committee on ethics to provide strict monitoring of these principles and avoid excesses in this area."[103] (Pompidou's perspective is particularly important, because he went on to become the president of the European Patent Office from 2004 to 2007.) But Gérard Caudron, a French member of the European Socialist Party, responded,

> There is no such committee in place. Who will set the limits in these circumstances? What is the precise definition of a human being? On the basis of the past, or looking at what is happening almost before our very eyes, we may well be tempted to think that there is no single definition. We are dealing here with matters where ethics must inevitably take priority over considerations of industrial competitiveness. Deciding to issue patents of ownership on life itself is an important step which cannot be based on considerations of "what is right for industry."[104]

At the same time that they debated the meaning of patents themselves, they struggled to identify the appropriate forms of governance—including their own roles. But parliamentarians were also distinguishing between what was right for society and "what is right for industry" in a manner unheard of in the United States.

These concerns led to forty-six amendments, which the Parliament passed along with the original Commission proposal in October 1992.[105] Passage of these amendments virtually ensured that the Commission would have to submit a revised BPD for approval by both the Parliament and by the Council of the European Union. This was because there were many contradictions between the BPD's text and the amendments, and because neither the Commission nor the Council would agree with all of the Parliament's amendments.[106]

This approach brought the Parliament into conflict with the Commission. This conflict would soon become even more meaningful, because in 1993 the Parliament would gain additional power in determining the outcome of the BPD. The Treaty of Maastricht, signed in 1992, strengthened the EU's governance structures, created the "euro" currency, and instituted the "co-decision" procedure, which meant that the European Parliament would now have to approve legislation in addition to the Council.[107]

As they delayed the BPD's passage and called into question life-form patentability, EU parliamentarians understood patents as having ethical, social, and economic implications and envisioned a patent system that would take responsibility for these issues. They also saw it as a deeply political question, to be answered through democratic deliberation among elected representatives rather than only among legal and scientific experts. Finally, many saw "ethics" experts, who had not played a role in any patent system previously, as relevant.

The Commission issued a revised proposal in December 1992. It interpreted the parliamentary discussion as focusing on the "ethical dimensions" (which it defined as focusing on humans and animals), ignoring most of the other concerns raised. It argued that these ethical concerns had been implicit in its previous draft, but acknowledged that this was not enough. It now argued that "patent law should contain certain impassable barriers so as to provide guidance for those interpreting the concepts of public policy or morality."[108] It interpreted the *ordre public* clause in terms of bioethical consideration, but it disagreed with many MEPs on where such barriers should be erected. It advocated the patentability of human genes and argued that patent law was not an appropriate place for monitoring the application or commercialization of research. It also noted that the European Union had developed numerous programs to study and address bioethical concerns, calling attention to the GAEIB, which it had asked to review the BPD. By 1993, the GAEIB had issued an opinion that largely agreed with the Commission's proposal, and by 1994 the Council of the European Union had issued its support of the proposal as well.[109] Pressure on the European Parliament to approve the BPD was growing.

But the Commission's changes did not have the force of law. They appeared mostly in "recitals" that preceded the BPD's text, which were not legally binding but were meant to provide future decision makers with interpretive tools. One of the recitals noted, "Whereas it is desirable to include in the body of the Directive such a reference to public policy and morality in order to highlight the fact that some applications of biotechnological inventions, by dint of their consequences or effects, are capable of offending against them . . ."[110] Beyond these recitals, the Commission erected ethical barriers in only rare and extreme conditions when compared to the amendments the Parliament had passed on the previously proposed BPD. It prohibited patents on processes of modifying the genetic identity of humans only if there was *no* therapeutic purpose, and on the processes of modifying the genetic identity of animals only if there was *no* benefit to man or animal.[111] It did not include a prohibition on human gene patents, which many parliamentarians had advocated.

The majority of MEPs were unhappy with the revised BPD, and a growing group of activists emboldened them. Led by Austrian environmental organization Global2000, the German group Kein Patent Auf Leben (No Patents on Life), the Green Party, and Greenpeace, opposition to the BPD included scientists, physicians, patient advocacy groups, development organizations, and environmentalists across Europe.[112] Everyone had his or her own reasons for opposing the BPD. Patient advocates and scientists worried about whether patents on human genes might slow research and increase the costs of health care.[113] Environmental groups and development organizations opposed the consolidation of the agricultural industry and the ecological implications of genetic engineering.[114] Almost all of them emphasized the moral significance of commodifying life.[115] Together, they sent letters to and visited MEPs, and held demonstrations in front of the European Parliament in Strasbourg.[116] They also worked directly with sympathetic MEPs, including Hiltrud Breyer and Benedikt Haerlin of the Green Party, to ensure that they were taken seriously. In sum, they were trying to open the debate back up to include a broader understanding of the implications of patents and their consideration in policy, which some of the parliamentary committees had advocated initially.

By the time the BPD came back to the Parliamentary floor in March 1995, these activists had had some impact. Parliamentarians incorporated not just arguments, but also images from opponents' campaigns, into their statements. Questioning the morality of commodifying life, for example, an Italian MEP who represented the Green Party observed,

> I think that our thanks should go to the Greenpeace activists who pointed out to us, even yesterday and today, what was at stake. They distributed a small picture which said all through the force of the visual image: a picture of paradise dating, I believe, from the seventeenth century, in which we see Adam and Eve and animals and plants. But it differed from paradise as we imagine it in this way: on every plant, tree and animal there was a little card bearing a company name, and we could even imagine a price ticket.[117]

Overall, parliamentarians were still unhappy with the BPD. The Commission had not addressed their initial concerns, and now European citizens and civil society groups were amplifying the issues they had raised initially. Whereas in the United States the Supreme Court and other amicus briefs had dismissed the PBC's arguments as irrelevant, in Europe policymakers made these points themselves. This was not simply the result of differences in the venues where the discussion took place. After all, the US Congress could have intervened if it wanted to do so. Rather, as we saw in chapter 1, the US and Europe had long histories of thinking quite differently about the relationships between governments and markets vis-à-vis innovation, and about the distinction between the inventor's and the public's interest in the patent system. While many of these differences had disappeared or lost their force throughout the twentieth century, in Europe they provided a basis for renewed calls for democracy around life-form patents and could take a somewhat new shape in the form of old legal provisions like the *ordre public* clause.

MEPs seemed most frustrated by the Commission's attempt to allow patents on human genes outside of the body. MEP Roberto Mezzaroma, of the center-right European People's Party, noted "It is unacceptable because it basically allows all biological materials to be patented. It is hypocritical first to claim that the genes, the proteins and the cells in their natural state belonging to the human body may not be patented and then maintain, incredibly brazenly and cynically, that all those elements that can no longer be ascribed to a particular individual—but whose structure may be identical to an element of the human body—cannot be excluded from patentability."[118]

Furthermore, Mezzaroma argued that a "subtle but very dangerous ambiguity pervades the whole of the text."[119] Specifically, he argued that the terms *public* and *morality* had not been sufficiently defined: "What is meant by 'public' and 'morality' in the context of our discussions? They seem to me to be too vague. And who is it that decides on the limits?"[120] These observations encapsulate well the underlying debate over the European patent system's political order. As they argued over whether, for example, human genes were patentable, they were also arguing over whose morality mattered, whether the "public" included all European citizens or just patrons of the innovation system, and how these decisions should be made.

In 1995, the Parliament defeated the Commission's revised BPD (in a 240–188 vote, with 23 abstentions). There had been major parliamentary opposition to the Commission's revision, and this opposition had come from members across political parties.[121] While the Green party was a major force against the directive, it had worked with Catholic parties from southern Europe to oppose it. Meanwhile, the parties who were major proponents of the directive (the European People's Party, the European Socialist Party, and the European Liberal Democrat and Reform Party) all had significant factions who opposed it. This kind of fracture may seem surprising, given the esoteric nature of the patent system and the tendency for parliamentarians to vote with their parties. But through the course of the debate, the BPD had become connected to emerging ideas about European democracy and morality, particularly in the context of growing public concern about the implications of biotechnology. Because it was one of the Parliament's first topics of discussion as it realized its new "co-decision" powers, it stimulated deep reflection that went beyond traditional party politics and created new concerns and coalitions.

The Commission Tries Again
Almost immediately, the Commission tried yet again with another proposal. This time, it incorporated many of CLACR's stricter amendments. It explicitly excluded from patentability all processes for cloning human beings, processes for modifying the germ-line genetic identity of human beings, and uses

of human embryos for industrial or commercial purposes. It also incorporated the balancing test developed by CLACR (and by the EPO, as we will see in the next chapter), weighing animal suffering against "substantial" medical benefit to man or animal, as well as the farmer's privilege. Finally, it acknowledged an advisory role for the GAEIB within the text of the directive itself. With these provisions, the commission had defined patents as influencing society beyond stimulating innovation and the market. It had also accepted an active role for government in the patent system and an explicit role for ethics expertise.

However, the new BPD still only addressed a subset of the concerns raised in Parliament. It allowed patents on human genes, for example, justifying them with the interpretation that had worked in the United States. The Commission argued that isolated human genes were "artificial" inventions rather than natural discoveries: "As regards the conventional principles of patent law, there is thus no difficulty in distinguishing between a discovery and an invention with reference to elements of human origin. Elements isolated from the human body by means of a technical process are *artificial* and thus qualify as inventions, since they are technical solutions invented by man in order to solve technical problems. *Nature is incapable of producing this type of element by itself.* The techniques employed in order to isolate such elements from the human body work only by means of human intervention [emphasis added]."[122]

Rather than focusing on issues of morality, and specifically human dignity and commodification, which had clearly motivated parliamentary concern, the Commission maintained its narrow technical orientation in its discussion of genes. Echoing the US approach, it simply dismissed ethical concerns and argued that in this case the issue at stake was the definition of nature and interpretation of the law, rather than the definition of life.

The European biotechnology industry mobilized to support the Commission's resubmission. It had remained relatively quiet during the first reading of the directive, perhaps assuming that the Commission would eventually prevail, since the issues were esoteric, technical, and legal, and since the Parliament lacked official authority. But this time it took the debate much more seriously. Representatives of patent lawyers, the biotechnology industry, and pharmaceutical companies lobbied MEPs through meetings and letters. The Institute of Professional Representatives before the European Patent Office (epi), for example, argued against the balancing test for animal patents in a letter: "It is very difficult for patent attorneys and Patent Office examiners to carry out the balancing required by this clause. Moreover, attitudes to benefits versus harm may vary from state to state. [The] epi therefore feels that this clause should be deleted as being impractical."[123] The industry also organized support from sympathetic patient advocacy groups. In one particularly vivid episode, lobbyists brought dozens of patient advocates in wheelchairs wearing shirts that said, "No Patents No Cures," to bear witness in front of the European Parliament's steps.[124] Accepting that the European Parliament approached the patent system's responsibility to the public interest in broader terms than GDP, jobs, and competitiveness, this event showed that the biotechnology industry was trying to demonstrate its health benefits for everyday European citizens.

Inside Parliament, response to the Commission's resubmission was mixed. CLACR and the Committee on Economic and Monetary Affairs and Industrial Policy supported the proposal, but the other committees that had written initial reports did not. The primary concern was still how the Commission approached issues of ethics in the context of patent law. MEP Breyer wrote that the proposed directive had "too restrictive an interpretation of the terms 'morality' and 'public policy,'" while MEP Friedrich-Wilhelm Graefe zu Baringdorf, a German member of the Green Party who drafted the opinion for the Committee on Agriculture, Fisheries, and Rural Development, worried about animal welfare issues, stating that "the concept of intellectual ownership of the 'design rights' of an animal presents many difficulties and leads to the presumption that animals are merely production machines or research tools to be redesigned and used for the convenience of humankind."[125] In addition, the Committee on the Environment, Public Health, and Consumer Protection argued that the ethics committee should be given additional powers, "to veto patent applications by qualified majority."[126]

But when parliamentary debate began again in July 1997, MEPs seemed less divided. A number of opponents continued to resist the "technical legal"[127] approach of the Commission, arguing that it had still not taken adequate account of the various implications of biotechnology patents or considered adequately the values and views of citizens. Some still worried that the directive certified the "commodification of life," while others felt that the European Union had abandoned its "obligations to the poorest part of the world" by allowing patents that might hurt small farmers.[128] Most MEPs, however, appeared satisfied with the language of the revised BPD and felt that it paid adequate attention to their concerns.[129]

They still saw the BPD and the patent system in moral terms, but their moral calculation had changed. MEP Roberto Mezzaroma of the European People's Party, who had fought against an earlier version of the BPD because it paid inadequate attention to "human dignity," was now convinced that it was necessary "for development in Europe and to create jobs."[130] Many were convinced by the patient advocates and industry lobbyists who had argued that without patents, research would proceed more slowly. Pat Cox, of the European Liberal, Democrat and Reform Party, observed, "Again today my mailbox is full of letters, not least from groups in my own country representing those who suffer from genetic and other medical disorders currently without a cure, requesting support for this measure because it offers them some hope. In conscience I will not vote against offering that hope."[131] By the end of 1997, the European Council had approved the BPD, and in May 1998, the European Parliament finally approved it as well. This was significant not only because of its impact on the patent system, but also because it was one of the first times that the European Parliament had played a significant role in developing European public policy.[132]

The final BPD passed by the Parliament and Council created a "farmer's privilege" and excluded patents on plants and animal varieties,[133] which minimized the impact of the BPD and maintained the existing plant-breeder rights system on agriculture. It also produced a new meaning of the *ordre public* clause, shaping the concepts of both "public" and "morality" in Europe with a focus on human and animal dignity, commodification of life, and bioethical concerns. Under that clause, it excluded the following from patentability: processes for cloning human beings; processes for modifying the germ-line genetic identity of human beings; and uses of human embryos for industrial and commercial purposes. It required a balancing test to determine whether processes for modifying the genetic identity of animals (and related products) should be patented. Suffering to the animal and damage to the environment would be weighed against "substantial medical benefit to man or animal." Finally, it acknowledged the advisory role of the Commission's European Group on Ethics in Science and New Technologies (formerly, the GAIEB).

The BPD did, however, allow patents on elements "isolated from the human body or otherwise produced by means of a technical process," including human genes. Although parliamentarians seemed to know at the time that this included patents on human genes, many would express confusion later about this provision. As we shall see in chapter 5, in the years afterward, many parliamentarians were strongly opposed to these types of patents—and said so in writing.

Overall, Europe produced guiding legislation that defined patents as having broad moral, social, and economic implications. This resurrected an approach to patents that went back to the early seventeenth century, but it also took into account new bioethical concerns related to commodification as well as human and animal dignity. With this definition came a system in which citizens and their elected representatives had an important role to play, and it had designated at least one new category of relevant experts, focused on ethics. While the European approach seems more open than the one in the United States, it too drew boundaries around what the patent system would consider and what it would not. And as politicians defined these boundaries, they used some strategies similar to those used by the Supreme Court and interests in the United States. They argued that human gene patents should be permitted, for example, by emphasizing the technical and legal dimensions of the domain and distinguishing it from ethical concerns. They also drew distinctions between simple and direct impacts on the one side and complex and attenuated ones on the other. This allowed them to justify interventions related to the commodification of life while arguing that they couldn't address most concerns related to farming and the environment. The latter issues, they suggested, were shaped by many areas of policy, of which the patent system was just one.

CONCLUSION

The prospect of life-form patents elicited debate in both the United States and Europe, but the issues were framed differently almost from the outset. In the United States, the discussion focused on matters of law and nature. Traditional organized interests and the Supreme Court dismissed the relevance of moral, social, and environmental questions, and even the idea that questions of life were at stake, by emphasizing the domain's narrow goal of certifying inventions. They distinguished between the patent system's impacts on the economy and innovation and the concerns Rifkin and his allies raised. Meanwhile, in treating *Chakrabarty* as though it was a customary patent case, the US patent system maintained the familiar political order: a transparent, evidence-based, and legally sound bureaucratic decision-making process with fair and objective oversight from the courts. Average citizens had little role to play in these specialized matters.

In Europe, by contrast, the life-form debate cast old concerns about the moral and social implications of patents in a new light, reflecting the rise of bioethics and concerns about biotechnology. Parliamentarians redefined the *ordre public* clause to focus on concerns regarding the commodification of life and the validation of controversial (and potentially unethical) areas of research. But they also addressed other social and environmental impacts beyond *ordre public*. While Europeans saw patents as having far-reaching implications, they restricted the BPD's scope—and the patent system's responsibility—by focusing on what they saw as the most direct effects. To them, this included moral impacts, which were defined in terms of human and animal dignity. Patents influenced the value of life, and the government needed to step in to protect it. This approach brought a new kind of politics. Parliamentarians and civil society groups had important roles to play, because they could introduce and evaluate citizen concerns. So too did newly emerging experts in bioethics. And a new decision-making tool, the weighing-up test, emerged that required a risk-benefit evaluation new to the patent system.

But these were only initial settlements. And controversy was growing on two related fronts.[134] Civil society groups around the world were arguing that patents on life-saving medicines gave their owners too much control and limited access to important inventions in socially damaging ways. There was also growing worry about the adequacy of regulation of medical and agricultural biotechnologies, including genetically modified foods, embryo research, genetic medicine, and reproductive technology, which posed moral, health, and environmental challenges. Central to both of these battles were questions about who controlled scientific and technological development, whether innovators were using their power in socially beneficial ways, and what role government should play.

3

COMMODIFICATION, ANIMAL DIGNITY, AND PATENT-SYSTEM PUBLICS

The *Diamond v. Chakrabarty* decision and the Biotech Patent Directive provided blueprints for how the US and European patent systems should approach life-form patents. Scientists then began to flood patent offices in both places with applications at the frontiers of genetics and biotechnology. But many questions were still unanswered. How would the United States and Europe handle, both legally and politically, the growing civil society concerns that patents could have negative moral and socioeconomic impacts? Would the PTO and other US patent-system institutions be able to maintain the characterization of their work as techno-legal and driving innovation? And how would the EPO and other European patent offices respond to the BPD, which defined patents as moral and socioeconomic, and outlined a more active role for government? Would continued pressure from citizen groups shift how legislators, patent offices, and the courts envisioned the responsibilities of the patent systems?

These questions emerged in the late 1980s as both the United States and Europe began to consider patent applications covering genetically engineered animals. Researchers had used recombinant DNA techniques to identify meaningful genes (e.g., linked to specific diseases, traits such as faster growth or more efficient digestion, or the production of pharmaceutically useful substances), insert them into animal embryos, implant the embryos into a female animal's womb, and then bring the fetuses to term.[1] The product of enormous scientific labor, the resulting transgenic animals could serve as models on which to test new medical treatments, such as those for cancer, to understand the consequences of particular gene mutations, and to track the progression of genetically caused diseases. But this genetic manipulation was an inefficient and time-consuming process, making the exclusive commercialization rights awarded to patent holders particularly valuable.

Social movements concerned about the environment, animal rights, and biotechnology on both sides of the Atlantic Ocean criticized the development of these genetically engineered animals, and they devoted much of their attention to patentability concerns. If no patents were permitted, their logic went, fewer genetically modified animals would be produced. They reiterated the worries articulated during the *Chakrabarty* and BPD debates that these patents would devalue animal life by turning it into a commodity, and that patents would encourage research in this ethically controversial area. As they made their arguments, they tried to claim new political ground. They initiated legislative hearings, tried to transform patent office processes into opportunities for public engagement, and launched new patent challenges in the courts.

The United States and Europe navigated these issues differently. They came to different decisions regarding animal patentability and the status of genetically modified animals as life forms in need of government protection. In the process, the United States reinforced its techno-legal definition of patents, while Europe maintained its moral and policy understanding. Meanwhile, decision makers and organized interests in the two jurisdictions did different kinds of political and institutional work—even inside patent offices—that support these understandings. This included producing different interpretations of appropriate participants and participation in the patent system: the United States deployed expertise barriers to maintain a narrow approach to its constituency, and Europe opened its doors much, much wider.

ASSERTING BARRIERS IN THE UNITED STATES

The PTO interpreted the *Chakrabarty* decision broadly. It soon began issuing patents on plants, and in 1987 it allowed a patent on its first animal: an oyster chemically altered to have three versions of each of its chromosomes.[2] Soon after this decision, the PTO's commissioner, Donald Quigg, made it official. He posted an announcement (which became known as the Quigg memo) in the PTO's official journal, the *Gazette*, explaining that the bureaucracy interpreted the *Chakrabarty* decision to allow patents on all "nonnaturally occurring nonhuman multicellular living organisms."[3] But "human beings" were excluded, the announcement stated, for legal reasons: they were prohibited by the US Constitution's Thirteenth Amendment banning slavery. There was no discussion of what constituted a human being for the purposes of patent examination, which would become a problem as biotechnology evolved.

The decision was criticized immediately. In a *New York Times* editorial, Jeremy Rifkin called the decision "radical" and "mindboggling," challenging the "non-elected bureaucrats in a Government agency, sealed off and isolated from public participation," who "have taken it upon themselves to reduce all living things to the new lowly status of 'manufactured processes.'"[4] He pressured Congress to intervene, wondering whether it would "meekly accept the edict handed down" or allow "ethical concerns" to triumph.

It made sense for Rifkin to turn to Congress. The courts and the PTO had already proven unsympathetic in the *Chakrabarty* case. Besides, he saw animal patentability as a policy question, requiring public participation and careful consideration of what he saw as ethical concerns, including the metaphysics of life. Indeed, Congress could still intervene in the patentability of life forms, although it had seemed unsympathetic to similar public-interest concerns in the past.

This pressure had an effect. Republican Senator Mark Hatfield of Oregon, the ranking Republican on the Senate Appropriations Committee, asked the PTO to delay issuing these patents so that Congress could study their "moral, political and scientific implications."[5] The agency complied. Then US Representative Robert Kastenmeier, a Democrat from Wisconsin who chaired the House Subcommittee on Courts, Civil Liberties, and the Administration of Justice, held four days of hearings on the matter between June and November of 1987, in both Washington, DC, and Madison, Wisconsin. Two senators from agriculturally rich states that engaged in animal breeding had opened up the moral questions regarding patentability again.

Asserting Rhetorical Boundaries

Kastenmeier invited thirty-three witnesses to testify during the hearings. An additional thirty individuals and organizations submitted written testimony. Unlike most previous patent-related Congressional hearings, many nontraditional participants—including environmentalists, animal-rights activists, bioethicists, religious leaders, and farmers—weighed in. Indeed, these hearings resembled the broad discussion developing around the European BPD at the time. As others have argued, opponents of animal patentability raised multiple moral concerns.[6] Some suggested, as Rifkin had in the PBC's *Chakrabarty* brief and many parliamentarians had argued in Europe, that animal patents would introduce a new kind of ownership that would erode animal dignity. John Hoyt, president of the Humane Society of the United States, an animal-rights organization, noted, "I submit, Mr. Chairman, that the patenting of animals reflects a human arrogance towards other living creatures that is contrary to the concept of the inherent sanctity of every unique being and the recognition of the ecological and spiritual inter-connectedness of all life. It also reflects a dehumanizing and materialistic attitude towards living beings that precludes a proper regard for their intrinsic nature. It suggests that animals have no inherent value other than that which serves the ends of human beings."[7] The secretary of the

National Council of Churches, which represented thirty-seven Christian faith groups from the Orthodox to Evangelicals, agreed: "The gift of life from God in all of its forms, and species, should not be regarded solely as if it were a chemical product, subject to genetic alteration and patents for economic benefits."[8]

Farmers raised a different moral concern, suggesting that animal patents would create innovation incentives in areas that promoted animal suffering. Speaking on behalf of a wide cross section of growers and breeders, Gervase Heffner, a lobbyist with the National Farmers' Organization of Wisconsin, noted, "The ethical question, the treatment of animals, I know that you have rules and regulations, but when I bred some of my own cattle, to a large Italian [bull], that some of those cattle went through excruciating pain and suffering when they had the offspring. No doubt about it, we have to be very careful what we are doing."[9] The primary concern of these farming groups was the genetic engineering of animals, and they suggested that patents had direct moral implications because they implicitly sanctioned certain areas of research and development that led to increased suffering.

Critics argued further that the moral impacts of life-form patents should trigger a different governance regime.[10] The usual procedures, which treated all technologies the same and relied on technical expertise and legal interpretation to make patentability judgments based on novelty, inventiveness, and descriptions, would not suffice. Some witnesses suggested a categorical exception for animal patents. Others noted that banning certain practices on the basis of moral concern was acceptable in other policy domains, and suggested that the patent system develop a similar approach. The representative from the National Council of Churches explained, "This background has led to legislation such as endangered species laws, animal welfare laws, laws regarding environmental quality . . ."[11] Still others advocated a broader approach to expertise in the patent system, including the incorporation of ethicists into the decision-making process. The Humane Society representative suggested the creation of a standing commission that would evaluate animal patents according to a case-by-case approach: "It seems to me that this is of such complexity and of such ethical and moral implications that probably some new commission or authority comparable to the Nuclear Regulatory Commission might have to be formed where people beyond scientists, people who come from the fields of morals and ethics, are a part of a commission to decide what is or what is not appropriate."[12] Because they saw patents as moral objects that shaped humanity's understanding and valuation of life, these interests suggested that religious figures and academics from the fields of philosophy and ethics had important insights to contribute. Their proposal echoed the approach taken by EU parliamentarians in the Biotech Patent Directive that would eventually pass in Europe, which articulated an explicit role for an ethics commission and identified some categorical exceptions.

But to the PTO officials, patent lawyers, and inventors, who were accustomed to a system that treated patents as purely techno-legal objects and to a patent system that was extremely limited in scope, these suggestions seemed simply ridiculous. As the Supreme Court had reinforced in *Diamond v. Chakrabarty*, patents served the public interest by stimulating innovation, nothing more. Animal-patent proponents suggested simultaneously that critics simply didn't understand the system. Dr. Alan Smith, a vice president at the US biotechnology company Integrated Genetics noted, "We are cognizant of the differences of opinion on these matters but submit that the ethical issues raised are not germane to the question of patenting animals. Indeed, we support the view put forward by William H. Duffey (Monsanto) in his testimony of the 21st of July that the patent system is certainly the *wrong place to regulate* matters of ethical, social, or moral concern [emphasis added]."[13] And Rene Tegtmeyer, the assistant commissioner of patents, stated simply, "Where safety, efficacy, environmental or other similar concerns arise, the use and sale of inventions are subject to laws other than the patent law. These are not patent law issues."[14] Facing organized pressure and scrutiny as never before, they argued that those concerned about the implications of the system were simply ignorant. As a result, they lacked the relevant knowledge to participate. They took an even stronger stance regarding the limited nature of the patent system than their counterparts earlier in the century had, perhaps because of the limited credentials of civil society representatives in comparison to those of the physicians and military officials who had intervened earlier.

Leroy Walters, a respected bioethicist at Georgetown University, also took the animal-patent proponents' view. He argued that critics' concerns should be "dismissed as either misguided or irrelevant."[15] He refused the idea that patents had any connection to animal welfare: "The ethical issues related to interspecies gene transfers or the patenting of animals should be distinguished analytically from the animal welfare questions. This type of analytical distinction is legitimate unless it can be shown that transgenic animal research or the patenting of animals necessarily or almost certainly will lead to the inhumane treatment of animals. I do not think the critics of either practice have made their case on this point."[16] Walters saw the relationship between patents and animal welfare as indirect, and therefore believed that the patent system was not the right place to consider such issues. In contrast, European parliamentarians had determined that patents on animals hurt their welfare directly, and therefore had limited the practice in their Biotech Patent Directive.

This rejection of moral concerns from a bioethicist would likely hold considerable weight among US decision makers. After all, it would be in the emerging profession's interest to suggest that bioethical expertise should be incorporated in a systematic way into patent examination.[17] Instead, Walters provided additional rhetorical ammunition against the charges that animal patents were unethical and that the patent system was responsible for animal welfare.

Patent proponents engaged only rarely with the substance of critics' arguments, preferring instead to note their inaccuracy and irrelevance. But when they did, proponents minimized the importance of such concerns. Monsanto's general patent counsel William Duffey, who appeared on behalf of the Industrial Biotechnology Association and Intellectual Property Owners, Inc., emphasized the patent system's economic benefits, arguing that the system "increases competition."[18] Patent lawyer Geoffrey Karny identified national economic risks as well as individual risks to companies that relied on their patent portfolios to raise capital: "Our country faces intense competition with other nations in the race to commercialized biotechnology. For example, it is well-known that Japan has targeted biotechnology for commercial dominance in the same way it targeted semi-conductors. Governmental action that unnecessarily or unreasonably hinders the development of biotechnology will make it more difficult for the United States to compete in world markets."[19] It was important to US international competitiveness, Karny suggested, to maintain the current approach and treat genetically modified animals as inventions like any other. This rhetoric would be particularly powerful in the 1980s, when the United States was concerned about the economic threat posed by Japan.[20] To the extent that values entered the discussion, then, they came in macroeconomic form. Stakeholders saw the patent system as working fairly and objectively to maintain a robust marketplace; it did not need to become embroiled in moral issues related to animal suffering and dignity.

Overall, these patent lawyers, companies, and representatives of the PTO, who all had traditional organizational interests in the patent system, performed a version of what sociologist Thomas Gieryn has called "boundary work."[21] Gieryn observes that in order to maintain their credibility and legitimacy in public debate, scientists use rhetorical resources to draw strict boundaries between what constitutes legitimate and objective science and what might be shaped by politics, morality, or religion. In the animal-patent episode, proponents understood patents as technical and amoral, and as lacking the implications that opponents described. The patent system itself had a technical orientation, grounded in scientific evidence and the law. This approach was associated with a fair, objective, and robust marketplace and economic growth. By contrast, the moral concerns of Rifkin and his allies were associated with uncertainty and politics, which if heeded would jeopardize not just the patent system but economic growth more generally.

The US patent system, as I described in chapter 1, had a long history of defining itself in technical terms and distancing itself from moral concerns. Its role was to operate fairly, objectively, systematically, and in a predictable manner, to ensure that inventors were rewarded appropriately and that the marketplace was healthy. Efforts throughout the nineteenth and twentieth centuries to consider moral and socioeconomic impacts had failed, and in *Chakrabarty* the Supreme Court had also distanced the patent system from the moral concerns raised by Rifkin and his allies. With this logic, if both patent governance and the marketplace functioned rationally and smoothly, moral judgments would result from collective, market-based decision making: good inventions would be plentiful, while bad ones would disappear due to lack of demand. The marketplace would produce the appropriate moral and socioeconomic implications of genetically modified animals. Europeans had also drawn boundaries around their patent system in the BPD debates. But, their boundaries looked quite different. US rhetoric distinguished between the technical and the moral, while organized interests in Europe had accepted the broad implications of patents but would argue they could only consider those that were direct. These differences reflected their respective understandings of the roles and responsibilities of government and the market.

Ultimately, after another round of congressional hearings that looked quite similar,[22] the patent proponents' strategy worked. Although both Representative Kastenmeier and Senator Hatfield proposed legislation to create a sui generis system for animal patents, neither was successful.[23] Congress lifted its moratorium in late 1987. Animal-patent challengers had managed to voice their alternative visions of patents and their governance through congressional hearings, but in April 1988 the PTO issued its first patent on an animal genetically engineered to contract cancer (known colloquially as "Oncomouse" because the research was done in mice).[24] The PTO's technical and limited understanding of the patent system's responsibility allowed it—at least temporarily—to reject the position of animal-patent critics.

Asserting Legal Barriers

Challengers tried again. This time, they used the courts, a frequent venue in the United States for addressing citizen grievances. By the 1980s, the courts were also becoming an important place to raise public-interest concerns, specifically to advocate on behalf of those with limited power and resources.[25] "Public-interest" lawyers were trying to create social and regulatory change using their legal expertise and the opportunities provided by legal statutes and case precedents.[26] Much of this legal work focused initially on the needs of underserved and unrepresented minorities, and by the 1970s it had extended to the consumer and environmental movements.[27] However, public-interest lawyers had almost no record of intervening in the patent system except in the *Chakrabarty* case just a few years before.

This time, the Animal Legal Defense Fund (ALDF) took the lead. Established in 1981 and based in California, its lawyers regularly used the courts to fight for animal welfare in research, farming, and hunting. It had helped to stop the importation of 71,500 monkeys for use by a private research company, blocked a US Department of Agriculture plan to brand dairy cows on the face using a hot iron, and stopped the US Navy from killing 1,500 wild burros.[28] It had become aware of animal patents soon after the congressional hearings and filed suit in a California district court on behalf of twenty-six plaintiffs, including the American Society for the Prevention of Cruelty to Animals, the Humane Society, the Wisconsin Family Farm Defense Fund, and the Association of Veterinarians for Animal Rights. Many of these plaintiffs had testified at the Kastenmeier congressional hearings and participated in the Rifkin coalition. (Rifkin's PBC was not part of this new coalition officially, but there were clear connections between the organizations.)[29]

In public statements announcing the lawsuit in 1988, plaintiffs echoed their earlier concerns, citing the "extraordinary economic, environmental, and ethical consequences" of animal patents.[30] John Kullberg, president of the ASPCA, noted that "the inhuman effect of [the Quigg memo] is to equate animals with light bulbs and pocket-sized fishing rods."[31] He suggested further that animal patents would "cause an increase in animal suffering and deformity."[32] To him, life-form patents had different implications from inanimate technologies, and the patent system was an important policy domain that had to address the additional implications of life-form patents in its decision making.

Perhaps anticipating that the courts would be unsympathetic to these concerns, given the fate of the *Chakrabarty* case and the Kastenmeier hearings, plaintiffs' arguments in the court proceedings focused on the legality of the Quigg memo. They argued that because it had not given public notice or sought public comment before issuing the memo, the PTO had violated the Administrative Procedure Act (APA). The APA, passed by Congress in 1949, required all bureaucratic rules to undergo a process of public notice and comment, to allow the public to participate in the rule-making process.[33]

The ALDF's decision to intervene is notable because it demonstrates a type of strategy that patent challengers in both the United States and Europe would use repeatedly. As we will see, challengers shifted their strategies to fit the venues available, but they tried simultaneously to transform these venues to address the concerns they were raising. This kind of transformative political strategy was necessary because patent systems had almost no record of citizen engagement. These groups had to create their own political opportunities.[34] In the case of animal patents, both the PTO and Congress had been unwilling to act. This left the courts, which had no record of sympathy to public-interest concerns as they related to the patent system. So the ALDF tried to use the APA as the vehicle for addressing their concerns. It could have chosen to fight on the basis of the "moral utility" doctrine that Justice Storey had articulated almost two centuries earlier. But, as discussed previously, the doctrine had fallen out of favor, and the plaintiffs must have predicted that they would stand a better chance by focusing on the potential APA violation.

However, the district court dismissed the case quickly. It argued that the Quigg memo merely interpreted the *Chakrabarty* decision, so the Administrative Procedure Act's notice and comment requirement did not apply. The ALDF appealed, and in 1991 the higher court, the Court of Appeals for the Federal Circuit, agreed with the District Court but added that the plaintiffs had no right to sue the PTO because they lacked legal standing to do so. This was a common difficulty for plaintiffs filing public-interest lawsuits. According to US law, to establish legal standing, they had to exhibit direct physical or economic harm that resulted from the "conduct complained of" and to demonstrate that a favorable decision would redress the injury. Both could be quite difficult in a case focused on social or environmental issues.[35]

Given its legal expertise, the ALDF had anticipated this hurdle and claimed that its plaintiffs would be harmed economically by the decision to allow animal patents. For the Wisconsin Family Farm Defense Fund, for example, animal patents would "increase [the] costs of operation by forcing them to purchase high priced 'patented' animals, or lose profits producing genetically inferior 'unpatented' animals; substantially increase livestock productivity and decrease herd size, causing a significant reduction in the number of farms and negatively impacting their communities . . ."[36] The ASPCA focused on the suffering that the patented, genetically engineered animals would endure: "Defendants' rule will cause an increase in animal suffering and physical deformity, forcing the ASPCA to increase its annual budget in order to provide care, medical attention, and placement services for the patented animals and will place further pressure on its humane law investigatory and enforcement staff."[37]

The Court of Appeals dismissed these standing claims. As it did so, it asserted legal boundaries that restricted who could participate in the patent system, to accompany the rhetorical ones drawn in the congressional hearings. At the same time, it reinforced the limited definition of patents themselves by denying implicitly the idea that patents caused animal suffering:

> Appellants baldly claim that they fall within the "zone of interests" addressed by the patent laws because patents "are issued not for private benefit but for the public good" and that "[p]atent case law emphasizes the importance of the public interest and the constitutional requirement of a public benefit." . . . In essence appellants' claim the patent statute's "zone of interests" encompasses any member of the public who perceives they will be harmed by an issued patent which they believe to be invalid. We cannot agree that the "zone of interests" of the patent laws is so broad. Under such an interpretation, we would, for example, be opening the door to collateral attack on the validity of issued patents; any competitor could simply file suit against the Commissioner challenging a patent's validity. This we decline to do. The structure of the Patent Act indicates that Congress intended only the remedies provided therein to ensure that the statutory objectives would be realized.[38]

The court decided that the plaintiffs lacked legal standing to sue because they did not fall within the "zone of interests" defined by previous patent cases. For the court, just as for the decision makers and organized interests who had weighed in on patents previously, patents were very limited technical and legal certifications that only had economic impacts on competitors. And even competitors could only challenge patents if they had suffered direct economic harm that could be ameliorated with a positive legal judgment. To the court, the concerns raised by the ALDF and its allies were too indirect to demonstrate "interest." Harms had to be clear and directly attributable to the patent, and even the economic costs articulated by the plaintiffs' were characterized as speculative and unrelated to patents. The court made this understanding explicit elsewhere in the decision, where it—like some witnesses in the congressional hearings—suggested that the relationship between patents and animal welfare was indirect and the responsibility of another policy domain: "For increased experimentation to lead to increased cruelty, appellants would have to allege that the existing animal cruelty laws are insufficient or that the issuance of 'animal' patents would 'encourage' researchers to disobey these laws."[39]

This reading of patents and of the law ensured that only other market players would be able to sue a patent holder. It also limited the kinds of concerns that might arise; market participants of course would be unlikely to challenge the moral status of patents, because they too might own patents. Although the plaintiffs appealed, the Supreme Court refused to hear the case. In the end, this court decision articulated rather narrow standards for who could initiate a legal challenge to a patent. In doing so, it made it difficult to shift the prevailing view of patents as legal, macroeconomic, and amoral objects.

Asserting Bureaucratic Barriers

Despite these setbacks, critics continued to raise concerns regarding the moral implications of patents. In 1997, Jeremy Rifkin tried another transformative strategy. Focusing on the lack of clarity in the definitions of "human being" and "non-human" in the Quigg memo, he decided to file a patent application on a human-animal chimera. He did so with the help of Stuart Newman, a developmental biologist from New York Medical College who had worked previously with Science for the People and the Council for Responsible Genetics, a biotechnology watchdog group. Their patent application covered a chimeric embryo comprised of both human and nonhuman animal cells.[40] What was the measure to distinguish between human and non-human, they asked? How many human cells did a chimera have to contain to be classified as and deemed unpatentable? It was a clever attempt. Regardless of how the PTO ruled on the patent, it would be defining the category of a human being. And so, particularly in an age of growing controversies over fetal tissue and embryo research,[41] Rifkin and Newman hoped to demonstrate both that patents had important moral status and that the bureaucracy played a significant role in shaping public morality. They also hoped to show that despite the PTO's assertions, it was making moral decisions with insufficient tools, and without proper public representation. The PTO examination process had never been used by civil society groups for such a purpose, although it had often been an initial testing ground for patentability before the courts or Congress considered matters further (the Chakrabarty patent provides an example of this kind of use). However, it was also a risky political strategy, since the PTO had already interpreted the Supreme Court's *Chakrabarty* decision broadly.[42]

As soon as they submitted the application in December 1997, Newman and

Rifkin held a press conference to announce the event and gave interviews to multiple media outlets. With this publicity, they defied the traditional conventions of the patent system again. Applicants generally keep their inventions quiet during the examination phase in order to prevent others from interfering with the process (for example, others might claim priority over an invention). But Newman and Rifkin wanted to encourage engagement among a broad set of citizens, not minimize it. In interviews with National Public Radio, *NBC News*, and the *New York Times*, they explained the reasons behind the application and argued that the patent system needed to recognize its moral significance and develop strategies to handle this responsibility appropriately.[43] Newman, for example, emphasized that his ideal approach would include the incorporation of explicit moral reasoning developed through deliberative democratic methods.[44]

Faced with this attention, the PTO took an unprecedented step and commented on the ongoing case. In a Media Advisory issued immediately after Newman and Rifkin began their promotional efforts, it stated that such a patent would not be granted because of the moral utility doctrine: "The courts have interpreted the utility requirement to exclude inventions deemed to be 'injurious to the well being, good policy, or good morals of society.'"[45] This was the first time that the PTO had invoked the moral utility requirement, either externally or internally, with regard to biotechnology. It had not figured in the filings in the *Chakrabarty* case or the congressional debates over animal patents, and it did not appear in the PTO's patent examination manual either. Indeed, as discussed previously, it had appeared rarely in American jurisprudence since its initial appearance in the early nineteenth century. Furthermore, it was unclear what moral utility meant in this context—would it be understood in cost/benefit terms, or did it refer to other modes of reasoning? Bruce Lehman, the PTO commissioner, stated in interviews that he would not allow patents on "monsters" or other "immoral" inventions.[46] This suggested not only that the PTO would determine what constituted a "moral" invention, but also that, rather than an approach that weighed costs and benefits, there might be an absolute prohibition on patenting certain inventions. However, this reasoning was not detailed any further (and Lehman himself, as I discuss below, was against an absolute prohibition).

As one might expect, the patent-law community rejected the PTO's language of moral responsibility strongly. It argued that it was not the PTO's job to be a moral arbiter and that it lacked the expertise to make such judgments.[47] Using arguments similar to those it had used in the *Chakrabarty* case and at Congress, it emphasized the patent system's role in certifying inventions and disseminating knowledge by making technical decisions simply based on novelty, inventiveness, nonobviousness, utility, and sufficient description. Lawrence Goffney, a former assistant commissioner for patents who had become a partner at a major patent-law firm in Washington, DC, observed that the PTO's job was "'to tell us it can be done.' Whether or not to do it is another matter altogether, and not the concern of the PTO. 'It's wrong to deny the patent on grounds other than the standard.'"[48] Goffney and other patent lawyers tried to reinforce the idea that the patent system was a narrow legal domain that could, and needed to be, kept separate from thorny moral debates that required broad public participation. As they did this, they rejected the idea that decisions about patentability were inherently moral decisions.

Not surprisingly, the PTO never actually used the "moral utility" doctrine in its examination of the chimera patent (nor did it ever discuss the doctrine again). Its personnel had used it solely to calm public fears. In its communications with the patent applicants, it also refused to engage with Newman and Rifkin's efforts to discuss the meaning of a "human being" for the purposes of patentability.[49] Deborah Crouch, the examiner in charge of the application, stated simply that the proposed invention included a human being and was therefore unpatentable as a result of the policy articulated in the Quigg memo. Rifkin and Newman jumped on her determination that the chimeric embryo was a "human being," and told the press that this proved that the PTO was making moral decisions.[50] Meanwhile, they tried to force her to articulate how she had defined "human being" so that it could be subjected to public scrutiny.[51] What criteria defined an organism as human, they asked? Who is making this decision and why do they have the authority to do so? Does the PTO have different definitions than the courts or the field of medicine? How does it distinguish between animal and human? With this interrogation, they charged that the question of what constituted a human being was not a technical but a policy one, which the public had to answer. As they did this, they sought to highlight the machinery and assumptions that lay underneath patents and the PTO's approach to governance and markets. Crouch responded to these questions by citing definitions of "human" from case law and from medical dictionaries.[52] Over the next eight years, the two parties communicated back and forth with incommensurable approaches to defining a human being, to patents, and to governance. As the applicants tried to draw out a more detailed understanding of the PTO's approach to a human being, Crouch simply restated the same handful of citations, treating it as though it were a clear technical object.[53] Newman and Rifkin finally abandoned the application in 2005.

As it emphasized the amoral status of patents, the PTO also maintained its technical approach to governance. When the case concluded, Commissioner Bruce Lehman remarked, "I do not believe there should be a prohibition against a human patent. . . . I was just deeply offended by anyone attempting to use the U.S. Patent Office to make a point, or to stop the advancement of science. I refused to make it easy for them."[54] To Lehman, the question of patenting a human being was scientific and legal, not moral. Therefore, there was no need to rethink which experts should be involved, how the patent should be examined, how the public should be engaged, or whether different branches of government should play a role. It was also unnecessary to consider the PTO's definition of a human being or its distinction between animals and humans. In fact, the idea that patents were moral objects was so incongruous with his approach that it was offensive.

New Institutional Arrangements

The chimera case was not a complete failure. First, it produced a new form of civil society engagement in the patent system. In the years afterwards, organizations began to seek other opportunities for public engagement that might lie within the PTO's apparently technical decision-making processes. In the 2000s, for example, the International Center for Technology Assessment (ICTA)—led by Andrew Kimbrell, a close colleague of Rifkin's—worked with the American Anti-Vivisection Society (AAVS), a group that had long campaigned against the use of animals in experimentation, to file "reexamination" requests on patents covering animals that had been engineered to serve as research tools but would likely experience significant suffering in the process. These included a rabbit whose eyes had been intentionally damaged to serve as an experimental model for dry eye disease, and an immunocompromised beagle that served to mimic humans who had weakened immune systems.[55] While the requests were clearly driven by ethical concerns,[56] the reexamination process, which became a part of US patent law in 1980, was extremely limited.[57] Third parties could only challenge patents on the basis of prior art, charging that the PTO had made some kind of error (including an omission) in its review. This focus on error is unsurprising, given the US patent system's focus on procedural objectivity in the examination process.

The AAVS and the ICTA tried to work within these constraints while maintaining their focus on ethics. In both the beagle and rabbit cases, they argued that the inventions were "obvious" in the context of prior art, and therefore unpatentable.[58] To make these claims, they relied on previously issued patents and scientific publications, forms of knowledge that were already established as legitimate in the domain. But they also argued that the animals were unpatentable on moral grounds. In the beagle case, they brought in the Supreme Court of Canada's 2002 ruling that higher-order animals were not patentable.[59] In the rabbit case, they suggested that animal patents violated the "moral utility" doctrine. To define morality, they relied on the US Animal Welfare Act, noting that "Congress explained that alternatives that replace the use of animals in experiments, such as product testing, should be actively encouraged and developed. Allowing patents for animals, however, provides financial incentives that encourage the development of animal models and this inhibits the growth of nonanimal alternatives."[60] The PTO dismissed the moral arguments quickly, as irrelevant to the reexamination process. But it revoked one patent on the basis of novelty and nonobviousness, and in another case the patent holders simply abandoned their invention by not paying PTO fees.[61] The ICTA and AAVS claimed success,[62] but they had not managed to convince the PTO to rethink its understanding of patents or their governance.

The chimera case also helped to stimulate a new organizational form within the PTO.[63] In the 1990s, the bureaucracy responded to growing scrutiny by creating the Sensitive Application Warning System (SAWS).[64] Its purpose was to provide officials with advance warning about patent applications that might generate widespread attention, so that they could both provide additional internal scrutiny and prepare for the publicity that might erupt. In other words, the PTO understood these controversies as creating public-relations problems. One of the officials who created the program noted, "Think of us as if we were some sort of corporation. The heads of a corporation are going to want to know if anything sensitive gets released out to the public. And it's no different for us. For the company may be worried about their stock or they're worried about their image. Well, we're worried about our image too."[65]

If an examiner found what she believed to be an application eligible for the SAWS, she was supposed to report it to her supervisor, who would recommend a second review. If the supervisor agreed with this assessment, she would report it to her superior at the next level. This process was supposed to continue until the report reached the offices of the Deputy Commissioner for Patent Operations and the Deputy Commissioner for Patent Examination Policy. Officials at this level would ensure that the examiner had properly conducted the review and quietly prepare for the possibility of heightened attention. In theory, after a SAWS review, the PTO might assess the public backlash and decide that it was too great to issue the patent. An official noted that in such cases, "the PTO will try to find some way to continue to reject the application. The PTO has lots and lots of tools. . . . So, essentially it is a question of finding a way to continue to reject it."[66]

In practice, however, the program had little importance even in terms of public relations. The SAWS was merely internal policy; it was not codified in the Manual of Patent Examination Practice, linked to statutes or case law, or discussed in official training sessions. Supervisors simply asked examiners to identify applications that could "generate high publicity or would potentially have a strong impact on the patent community" during the review process.[67] Although each department provided its examiners with some examples of the kinds of SAWS subject matter they might encounter, supervisors would rely mostly on examiners' own judgments to assess and identify which kinds of patents might fit the designation. They also emphasized that the SAWS did not influence the technical and legal determinations of patentability. One examiner explained the circumstances that might lead to a SAWS designation: "OK, if I am an examiner and I am working on a case and I see OK I've got the next [smartphone] application I am working on, if I am a primary examiner, I will want to at least inform my supervisor and inform my director that 'Hey I've got something here, it's going to be allowable, it's an allow form but it's high profile.' So, the SAWS is there to just identify cases that are to that extent."[68]

Multiple personnel at the PTO downplayed the importance of the program, and many examiners I interviewed did not even know of its existence.[69] One examiner noted that in her department, examiners reported applications that were candidates for the SAWS designation in the form of light-hearted forwarded e-mails; if an examiner saw an application that she deemed strange, funny, or particularly interesting, she would forward it to her colleagues, and her supervisor would decide whether the application was worth pursuing.[70] Perhaps examiners either dismissed the SAWS or transformed it into a type of comic relief because of the dominance of the rule-following culture at the PTO that I discussed in chapter 1. Because the system was never formalized, and because of the emphasis on detached application review, it was never clear how examiners were supposed to incorporate it into their decision making. In sum, while the SAWS was meant to identify proposed inventions that would capture public attention or controversy, the agency's emphasis on scientific and legal knowledge as a means of ensuring objectivity and detachment made its implementation unlikely. It required examiners to consider the social reactions to inventions with few conceptual tools to guide them—except for their own individual understandings of the world.

My own initial efforts to learn about the SAWS program demonstrate its inherent contradictions. In 2007 I filed a Freedom of Information Act request for documents related to the program. The PTO refused my request, arguing that it was part of the bureaucracy's technical procedures. "The information is a purely internal implementation of application screening procedures, and the release of some information contained in the internal records could be used to circumvent internal processes and practices, e.g., fashioning the language of a patent application to purposefully circumvent the screening procedures, if not render them entirely useless. Since these processes do not alter any standard of review applied to the applications, they do not affect the public, and therefore are considered purely internal, and not suitable for public disclosure."[71] In this communication, the PTO minimized the importance of its SAWS program by arguing that it was part of its regular review procedures. It also characterized this sensitivity determination as a technical process, reinforcing the idea that it did not engage in political or moral activities. Therefore, it was of no interest to the public.

The PTO dismantled the SAWS program in 2015.[72] In the preceding years, the program had become public and the subject of a lawsuit.[73] Patent lawyers and inventors had been furious, responding as they had to the invocation of the moral utility doctrine in the chimera case.[74] They argued that the SAWS introduced politics and uncertainty into an otherwise objective system. And because the PTO had never managed to make it systematic, or determined how to integrate it into its examination process, it abandoned the program. But the SAWS never really had a chance, because both patent-system institutions and their organized interests held so tightly to the idea that patents, and therefore patent governance, was techno-legal and therefore amoral.

Rather than provoking reevaluation of the moral implications of patents or their governance, the animal-patent debates led US stakeholders and decision makers to articulate stronger legal, rhetorical, and bureaucratic boundaries between the patent system on the one hand and morality and politics on the other. Using a variety of strategies, they simply rejected the idea that patents had the implications that critics described. As they did so, they produced an even stronger definition of patents as objective objects that warranted a highly technical approach to their governance. They also reiterated the idea that genetically modified animals were like any other technology, and that the fairest and best course of action was to treat them as such. Even the rare moments when the system sought to respond to growing scrutiny—the invocation of a moral utility doctrine in the chimera case and the SAWS system—seemed more like disconnected efforts in public relations. There seemed to be no interest among the PTO and traditional patent stakeholders in considering whether or how the system engaged in moral or political work. This fit with the idea that both markets and market making were rational and objective.

OPENING UP THE EUROPEAN PATENT BUREAUCRACY

The European debate over animal patents also began in the mid-1980s and proceeded simultaneously in two venues. Just as the European Parliament began to consider the BPD, the EPO received the "Oncomouse" patent application covering animals genetically engineered to contract cancer. The United States would grant this patent, filed by Harvard University scientists, in 1988.[75] But in Europe, this patent application, and the issue of animal patentability generally, stimulated two decades of controversy that led the patent system to rethink its definition of morality, its engagement with the public, and its understandings of relevant knowledge and expertise.

Considering New Publics

The Oncomouse episode proceeded differently from the European Patent Office's customary approach to patent review almost from the outset. The first group of technical examiners rejected the application in 1988, citing the 1973 European Patent Convention's exclusion of patents on "animal varieties" and invoking the *ordre public* clause.[76] They observed that "the idea of the patenting of higher organisms has encountered severe criticism for ethical and economic reasons."[77] But they also suggested that the legislature had a role to play in this discussion,[78] citing both the US congressional hearings, which were going on at the time, as well as the proposed BPD. The applicants appealed, as they commonly did in patent cases, and a larger group of examiners reversed the initial decision. They decided that patent law was "not the right legislative tool" to deal with ethical concerns, but rejected the patent on other grounds.[79] This decision was made, after all, before the Parliament had approved a revised BPD. But the applicants appealed again, this time to the Technical Board of Appeal (the EPO's penultimate appeals board, made up mostly of technical examiners). As the TBA examined the case, the EPO would experience pressure related to how it defined morality and how it defined appropriate participation in its decision making that was far greater than it had ever experienced before.

Citizens with interests similar to those of the US animal-patent challengers—farmers, religious figures, animal-rights activists, environmentalists, scientists, veterinarians, and members of the Green and Socialist Parties—submitted hundreds of pages of observations to the TBA hearing. All raised concerns about genetic engineering research and about commodification.

They urged the EPO to reject the Oncomouse patent because it was "incompatible with any respect for life"; it defied "the basic values of our democratic society"; it demanded "the growing scientific exploitation of the helpless"; it would lead "to the patenting of human beings, our children, our body parts and our souls"; and it ensured that "the animal world will be degraded as a source of objects whose only purpose on this earth is to be manipulated for short-term profit [translated from German]."[80] To them, as to their counterparts in the United States, animal patents had profound negative implications for public morality.

Although the flood of third-party submissions in this case was unusual, the EPO allowed anyone to submit such "observations" in its proceedings. The EPO was under no obligation to read or consider these concerns. Complicating matters, however, was the fact that parliamentarians had begun to question the moral dimensions of the BPD by this time. So, the EPO would be making its decision in an increasingly contentious political atmosphere and under great scrutiny. This explains, perhaps, why the TBA considered the third-party submissions and brought the *ordre public* clause back in its 1990 decision:

> The genetic manipulation of mammalian animals is undeniably problematical in various respects, particularly where activated oncogenes are inserted to make an animal abnormally sensitive to carcinogenic substances and stimuli and consequently prone to develop tumours, which necessarily cause suffering. There is also a danger that genetically manipulated animals, if released into the environment, might entail unforeseeable and irreversible adverse effects. *Misgivings and fears of this kind have been expressed by a number of persons who have filed observations with the Board,* . . . [and] considerations of precisely this kind have also led a number of Contracting States to impose legislative control on genetic engineering. The decision as to whether or not [the *ordre public* clause] is a bar to patenting the present invention would seem to depend mainly on a careful weighing up of the suffering of animals and possible risks to the environment on the one hand, and the invention's usefulness to mankind on the other. It is the task of the department of first instance to consider these matters in the context of its resumed examination of the case [emphasis added]."[81]

The appeals board could have characterized patents and their governance in narrow and techno-legal terms, as the PTO had done under similar circumstances. It could have interpreted the *ordre public* clause as applying only in rare circumstances, as it had done previously. It could have rejected the connection between patents and animal welfare, as the PTO and many traditional stakeholders had done in the United States. Instead, it too defined patents as potentially harming public morality by encouraging the commodification of genetically modified animals. It also suggested that the substance of the invention mattered in the decision to grant a patent, and it drew connections between patents and animal suffering rather than leaving animal welfare issues to other policy arenas (suggesting that patents increased research, which in turn increased animal suffering). It also took responsibility for these consequences and extended the reach of the *ordre public* clause by validating the weighing-up test that would eventually make its way into the BPD. (It is unclear whether MEP Willi Rothley or the EPO was the architect of the weighing-up test, but there is evidence that they were coordinating with one another, as the idea appeared at the EPO and in the European Parliament around the same time.)

Finally, we can see that the TBA acknowledged the "misgivings and fears" expressed by the "number of persons" who submitted comments to it, suggesting that public concern was relevant to its decision making. This acknowledgement is surprising, particularly in comparative perspective. Unlike the PTO, which defended its expert space, the EPO not only allowed public input but mentioned it in its consideration of the case. This suggested that, at the very least, it shaped the decision-making environment. The EPO thus implicitly validated the participation of new kinds of third parties in the patent system.

The TBA sent the Oncomouse application back to examiners, who performed the weighing-up test for the first time. The examiners decided that the benefits outweighed the risks, and in May 1992 the EPO issued the patent covering a transgenic, nonhuman mammal engineered with an oncogene sequence, and the methods of making it.[82] This was the same version of the Oncomouse patent as the one that the US PTO had issued in 1988. In other words, the EPO's apparent consideration of public concern and its incorporation of morally explicit decision making didn't seem to make much of a difference in the decision.

This decision aligned with others that the EPO was making at the time. As it navigated the Oncomouse case, the EPO was also dealing with civil society criticism of its decisions allowing patents on genetically modified plants.[83] Greenpeace and the Green Party, among others, had begun to use the EPC's opposition mechanism to challenge patents they deemed morally and social problematic. The mechanism allowed third parties to "oppose" a patent on any grounds, including *ordre public*, within nine months of its issue. Competitors customarily used the opposition mechanism to challenge a patent's validity.[84] But these groups were trying to transform the mechanism, as Rifkin and his colleagues had tried to do with the examination and reexamination processes in the United States, into a political opportunity to force attention to their concerns.

In these early efforts, the EPO responded with narrow interpretations of the *ordre public* clause. It argued that the clause was relevant in plant-patent cases only if the invention was "likely to seriously prejudice the environment or [was] contrary to the conventionally accepted standards of conduct of European culture."[85] This "threat" had to be "sufficiently substantiated" at the time the EPO decided to revoke the patent. The EPO also repeatedly rejected the evidence and expertise opponents presented—including public-opinion data, surveys, and national laws—arguing that they were insufficient to prove violation of the *ordre public* clause. But even in these interpretations, EPO personnel did not reject a link between patents, innovation, and the environment and public morality, as their US counterparts usually did. Rather, they simply established a very high threshold for patents whose implications were so problematic that they required intervention. They would soon reevaluate that threshold as well.

Reinterpreting Ordre Public

Despite the EPO's 1992 decision, the Oncomouse patent controversy was just getting started. Twenty-one groups—including animal-rights organizations, religious associations, environmentalists, the Green Party, and even the German government's Ministry for Youth, Family, and Health—filed official oppositions against the Oncomouse patent.[86] They claimed that the patent was issued in error because it violated the *ordre public* clause. It created, they argued, "a new form of animal ownership which carries no duty of care."[87] They also challenged the weighing-up test, both in terms of how the EPO had performed it and whether it was an appropriate method of evaluation. They focused on procedural error, just as civil society groups had done in the United States with the reexamination mechanism. But here the evaluation (and the error) involved an ethical assessment, which could create an opening for non-technical knowledge and expertise.

As with the third-party observations filed in the earlier phases of the case, the EPO had to decide how to handle opposition. The opponents had followed the minimal legal requirements and paid the relevant fees, but they had different interests and concerns from the industrial competitors who usually participated in EPO procedures. Unlike the United States, where strict guidelines at the PTO and in the courts shaped who could participate and how, the EPO chose to treat the new participants the same as any other third party, and

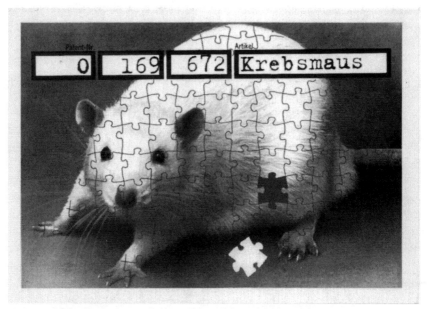

Figure 3.1. Front side of postcard sent from citizens to German legislators in response to the EPO's Oncomouse patent, 2004.

began an official opposition process. At a time of growing public scrutiny, the opposition mechanism was becoming an acceptable avenue for public participation in the pan-European patent system.

Over the next two years, in preparation for oral hearings in November 1995, the opponents, the patent holders, and the EPO exchanged hundreds of communications. The patent holder filed documents defending the Oncomouse's patentability. The opponents offered their own evidence for revoking the patents, including that they represented a large, critical public. They submitted petitions opposing the patent, supported by hundreds of thousands of signatures from European citizens. There were critical letters from citizens directly to the EPO. Opponents also sent the EPO copies of hundreds of postcards that German citizens had sent to members of the German Parliament, asking for their help in opposing the Oncomouse patent (figures 3.1 and 3.2). One side of the postcard showed a picture of a mouse divided into puzzle pieces, with one piece altered to signify genetic modification. The other side provided space for the recipient's name and address, and included the following text: "Stop the abuse of patent law now. Ensure that patents are confined to real inventions. Stand up for renegotiation of the EU Biotech Patent Directive. Show

Sehr geehrtes Mitglied des Deutschen Bundestags,

mit der „Krebsmaus" wurde 1992 zum ersten Mal ein Patent auf Säugetiere erteilt und Leben zur menschlichen Erfindung erklärt. Inzwischen gibt es massenhaft Patente auf Saatgut und Tiere, auf Gene und sogar auf menschliche Organe und menschliche Embryonen. Diese Patente werden gerechtfertigt mit der EU-Richtlinie 98/44, die jetzt auch in deutsches Recht umgesetzt werden soll. Stoppen Sie jetzt den Missbrauch des Patentrechtes. Sorgen Sie dafür, dass Patente auf wirkliche Erfindungen beschränkt werden. Setzen Sie sich für eine Neuverhandlung der EU-Richtlinie ein. Zeigen Sie Respekt vor dem Leben. Das Patentamt wird sonst den Unterschied zwischen Menschen, Mäusen und Maschinen nie begreifen.

Mit freundlichen Grüßen

Platz der Republik 1
11011 Berlin

Vom 5.–9. Juli 2004 findet am Europäischen Patentamt die entscheidende und letzte öffentliche Verhandlung zum Patent auf die „Krebsmaus" statt.

* Setzen Sie den Namen Ihrer/s Bundestags-Abgeordneten oder einen der folgenden Namen ein:
Dr. Maria Böhmer (CDU), Horst Seehofer (CSU),
Dr. Ernst Ulrich von Weizsäcker (SPD),
Dr. Reinhard Loske (B90/GR)

Figure 3.2. Back side of postcard from citizens sent to German legislators in response to the EPO's Oncomouse patent, 2004.

respect for life. Otherwise, the patent office will not understand the difference between men, mice, and machines [translated from German]." Like Jeremy Rifkin and Stuart Newman in the United States, these opponents sought to stimulate outrage and alert the EPO to this engaged citizenry.

In explaining their grounds for opposition, challengers made arguments and submitted evidence that was quite different than what the EPO usually saw. First, they challenged the appropriateness of the weighing-up test in evaluating the moral implications of the patent. Instead, they advocated a deontological approach that considered, in and of itself, the morality of both commodifying animals and making them suffer. To justify this argument, they relied on reports from medical ethics committees and religious experts.[88] They also suggested that the best way to assess *ordre public* was through public opinion. If the EPO had conducted such an analysis, opponents noted, it would have found clear opposition to the patent. They supported this claim in several ways. They referred to the individuals who signed their petitions and sent letters and postcards to the EPO.[89] They provided survey evidence from the widely respected Eurobarometer survey,[90] performed by the European Commission on a wide variety of issues related to the European Union. They referred to a resolution passed by the European Parliament in 1993, requesting the Onco-mouse patent's revocation and a temporary moratorium on all animal patents until parliament had resolved the BPD, since by then the EPO had agreed to comply voluntarily.[91] The EPO would have to decide whether and how to consider these nontraditional forms of evidence, which US patent-system institutions had largely decided were outside their scope.

Second, opponents argued that the EPO had conducted the weighing-up test incorrectly. Citing articles in leading scientific and medical journals (and many published by the patent applicants themselves), which documented the extreme pain experienced by animals suffering from spontaneous tumors, they claimed that the patent bureaucracy had not adequately evaluated the suffering that the animal would endure.[92] They supplemented these articles with photographs of the suffering animals. Scientific articles were a familiar source of evidence in the patent bureaucracy, but opponents were using them in a new way. They were asking the EPO to consider the ethics of the research methods used, rather than to treat the content as evidence of "prior art."

They also challenged the "benefit" side of the test, arguing that the EPO and patent applicants had overstated the potential for benefit from this research, because rodent tumors do not mimic human tumors.[93] One article that they used in support of this assertion, from the *British Journal of Surgery*, stated, "Unfortunately biological behaviour and chemotherapeutic sensitivities [of mice] do not closely resemble those found in human solid tumors. Chemo-therapeutic data derived from these experimental systems may therefore be misleading with the result that patients in clinical trials frequently receive in-effective agents."[94] Finally, opponents argued that the EPO had not considered alternative methods for testing chemotherapeutics that would result in equal benefit for cancer patients but lower levels of animal suffering.[95] Both Britain and the United States, they observed, had reduced animal use consider-ably in their research efforts and had begun to develop alternative models. Furthermore, most animal-welfare laws encouraged the development of alter-native models for experimentation. Scientists at the University of Cambridge, for example, had "developed an automated cell system for screening new anti-cancer drugs."[96] The availability of these alternatives, they concluded, should affect the assessment of the invention's benefits.

Overall, while the EPO and the opponents seemed to agree that animal patents had moral implications, opponents questioned the scope of the EPO's ethical evaluation. They argued that the EPO needed to put the invention into broader scientific, medical, and social context, and use what they deemed ap-propriate expertise to do this type of assessment. They also argued, this assessment required a new kind of public involvement. The public's role was no longer to gaze upon the techno-legal proceedings and trust the results. It had the right to influence patent decision making directly.

Despite these unusual submissions, the EPO's only initial change to its process was to prepare for an unusually large crowd at the oral hearings. Usually, the proceedings lasted one day, were fairly small, and included a handful of spectators. But the EPO predicted that there would be a much larger audience. It could have limited the audience to the seating available in the hearing room and turned citizens away, but it seemed to calculate that it needed to demonstrate its transparency and the objectivity of its processes to a large audience, given the public attention. So it changed the seating arrangements in the main meeting room to accommodate more spectators and created an overflow room for members of the public to watch the proceedings on closed-circuit TV.[97]

On the day of the hearing, it became clear quickly that, indeed, the proceedings would be unlike anything the EPO had seen before. It was overwhelmed by people who wanted to attend. And these people were not silent. At one point, the head of the Opposition Division suggested to the patentee's lawyer how he could amend the patent so that it could be maintained.[98] This was common practice in an opposition hearing, but the opponents and audience members saw it as demonstration of the EPO's bias toward the patentee, and more generally toward industry. They began to boo and hiss. A matter that the EPO Opposition Division saw as technical was seen by the opponents' supporters as deeply social and political. To the opponents, treating patents as moral objects required the EPO to do more than just perform the weighing-up test. They needed to reevaluate a variety of mundane practices such as these.

As the hearing continued, the Opposition Division struggled to deal with the new participants, both the official opponents and members of the audience. After an unusually lengthy four days, the EPO suddenly ended the proceedings. It told the opponents and patentees that further discussion could proceed in writing, a turn of events that opponents also saw as biased toward the patentee.[99] Although it is unclear why the Opposition Division concluded the hearing so abruptly, it is possible that it was too uncomfortable with the pressure coming from the unusually large and vocal audience.

These hearings demonstrated the depth of the incommensurability between the opponents and the EPO, as the chimera patent application process had in the United States. But, as it failed to do at the PTO, the episode ultimately inspired a change at the EPO in terms of how it understood its role and its decision making. A member of EPO's communication department who was advising the Opposition Division through this episode recalled that when the OncoMouse Harvard patent was granted in 1992, a big storm arose that nearly swept away the EPO, because people were so strongly inward-looking at this place that they didn't realize what was going on. That the work that they were doing was not just considered to be of relevance to pharmaceutical companies and other innovators, but there was a strongly social component to the patents as well. . . . I told [the Opposition Division] straight away: 'This is not an opposition hearing, a normal one [in which] you get inter-parties procedure. This is going to be turned into a tribunal against the EPO. . . . What you are doing is wrong. You are not paying attention to the people. You're not seeing the political and social dimension of this patenting process. This patent is a contract between a private investor, someone who's got a private interest in obtaining a patent, and society. You are mediating this, and you're in the middle of it. And if you're not doing the job properly, you're going to get some feedback.'"[100]

Previously the EPO's understanding of the system had been similar to that of the PTO. Its public was its inventors, and it was charged to mediate among innovators in an amoral manner, using tools from science and the law. But the Oncomouse case caused patent officials to think more explicitly about patents as moral objects, about the possibility that average citizens might have legitimate interests in the system that might be quite different from those of innovators, and about themselves as balancing public and private interests. Of course, EPO officials were somewhat more predisposed to this position than their US counterparts. Given the market-shaping legacy in many European countries, there was some strength in the idea that government should represent the public interest and play a mediating role as it shaped the objects that received patent protection in the marketplace. And there was a legacy of this in its patent system, in which the *ordre public* clause was a remnant.

Over the next six years, the EPO continued to receive observations from the patentee and the opponents, but waited to reach a decision until the European Parliament and Council passed the Biotech Patent Directive. Once the BPD affirmed the weighing-up test in dealing with animal patents, the Opposition Division held oral hearings again in 2001 to bring the Oncomouse matter to a conclusion. In January 2003, the Opposition Division approved a substantially narrower patent, covering only rodents engineered to have the oncogene. It argued that under the narrower conditions the application passed the weighing-up test that had been codified in European law.

The patent applicants appealed again to the TBA. In a decision that must have frustrated the applicants, the TBA then narrowed the patent further still.

It decided that in order to pass the weighing-up test, the patent could only cover *mice* genetically engineered with the oncogene sequence.

> There is no evidence on file, either in the patent itself or elsewhere, that any such benefit, let alone a substantial medical benefit, is likely to be derived from applying the claimed process to all rodents, or indeed to any animals of the order *Rodentia* apart from mice. . . . There is quite simply no evidence to show that all the various animals in the category of rodents are so different that each of them would provide a contribution to cancer studies, such as being specifically suited as a model for studying specific types of cancer."[101]

The applicants' research had focused only on *mice*, and therefore they could only prove that the mouse invention would "likely" have substantial medical benefit. The TBA approved an amended patent covering only mice, and the higher board of appeal, the EBA, did not take the case.

In the end, the EPO allowed a patent, although it was far narrower than the original Oncomouse application as well as the patent that had been granted in the United States. And as it did this, it took the opponents seriously. In the context of the European public's attention to the Oncomouse case and both life-form patents and biotechnology more generally, the recently passed BPD, and its own exposed and unclear status between national governments and the European Union, it chose to engage with citizen contributions by issuing a narrower patent. This move certified public-interest groups as legitimate participants in the European patent bureaucracy, in contrast to their US counterparts. And it validated the opposition mechanism and appeals processes as avenues for public engagement.

In this process, the bureaucracy defined patents in explicitly moral terms. It saw patents as potentially encouraging the exploitation of immoral inventions, and its role was to assess "whether or not the exploitation of the invention conformed with conventionally accepted standards of conduct in European society."[102] To do this in the Oncomouse case, it had to balance animal suffering and benefit to humankind. Thus, the EPO validated a connection between patents and animal welfare, which US decision makers and stakeholders had rejected, and established its own authority over these matters.

But the EPO had also decided *how* it would evaluate moral concerns. The weighing-up test allowed it to focus its approach on animal suffering and the benefit to humankind, but the EPO still had to choose what evidence and expertise to consider. In this case, it demonstrated a clear preference for abstract legal reasoning. It largely rejected opponents' empirical contributions on the

weighing-up test by arguing that examiners only needed to consider information about *likely* animal suffering, *likely* substantial medical benefit, and the
correspondence between the two, claiming that "other matters such as the degree of suffering or the availability of non-animal alternatives need not be considered."[103] To the extent that it would consider empirical evidence, it would
only focus on what was known on or before the date of filing the application.
Of course, such evidence would be virtually impossible to produce: very few
people other than the patent applicant would have conducted research on
the benefits and risks of the proposed invention before the filing date for the
patent application, and the patent applicant would be unlikely to publicize
such information unless it demonstrated a clear benefit.

The TBA made a similar argument against opinion polls, concluding that
it "did not consider opinion polls reliable tools for assessing public perception."[104] Without extensive information about each poll's methodology, it
could not judge its utility. But even if it had this methodological information,
the TBA suggested, it would be impossible to know "whether the respondents
were stopped on street corners and answered questions in a hurry or were invited into comfortable premises and given time to think."[105] Although the TBA
did not reject a responsibility to evaluate empirical evidence, its responses
suggested both skepticism and discomfort. Its personnel seemed much more
willing to make judgments on the basis of abstract analysis, likely because of
their legal backgrounds and training.

Incorporating Ethics into Everyday Bureaucratic Practice
As the Oncomouse case proceeded, the EPO developed multiple institutional
initiatives that reinforced its new and expanded understanding of its role, its
public, and its approach to relevant knowledge and expertise. Many of them
were designed to change bureaucratic culture, including the daily work of examiners and other EPO personnel. Beginning in the early 1990s, for example,
the EPO's internal newsletter started to discuss the emerging life-form patent
concerns in multiple ways. Called the *Gazette* like the PTO's version, and dis-
tributed to all employees, the newsletters reprinted relevant EPO press releases announcing the grassroots oppositions and displayed photographs
of the activist protests occurring outside of its buildings (figure 3.3).[106] Previously, the *Gazette* had focused mostly on internal announcements, including listings of job openings, promotions, and new employees; short articles
about foreign visitors to the EPO; and announcements about and reviews of
(mostly social) EPO events. Then in 1992, the *Gazette* announced an essay competition on "Patents and Ethics in the Context of Modern Technology" among

Activists demonstrating in front of the Isar building, against patents on life, before the Harvard Oncomouse hearing on 6 November

Figure 3.3. Image accompanying the *Gazette* article entitled "Harvard Oncomouse Opposition" (December 2001).

employees at the EPO. Two winners were awarded 15,000 and 10,000 German Marks (about 10,000 and 6,500 US dollars, respectively), and the essays were published not only in the *Gazette* but also in major European newspapers (*Financial Times, Le Monde, Neue Züricher Zeitung, and Süddeutsche Zeitung*).[107] With this attention, the EPO sanctioned attention to the growing controversy over patents, emphasizing its relevance to employees and encouraging them to form opinions. Officials could have decided that coverage of these issues would be distracting, demoralizing, or—perhaps worst of all—interrupt the objectivity of their employees' work. One could easily imagine that at the PTO, officials would be reluctant to educate examiners about the growing topics of controversy. However, the EPO's officials decided the opposite: thinking more about the issues of controversy, high-level officials seemed to believe, would make all employees do a better job.

These efforts continued, and were enhanced, through the 2000s. When Alain Pompidou, a biomedical scientist and doctor, became EPO president in 2004, he asked the EPO's Learning and Development Directorate (which is in charge of all personnel training efforts) to organize a series of lectures entitled "Ethics and Science: Are They Connected?" Pompidou had long-term interests in bioethical issues and had participated in science policymaking for years, including time spent as a European parliamentarian during the controversy over the BPD. The purpose of the nine ethics lectures, which were simulcast in the EPO's Munich and Hague offices and held over the course of a year and a half, was to "increase staff awareness of the ethical aspects of novel technologies."[108] Pompidou suggested the list of speakers for the series, most of whom were European professors of philosophy or bioethics. He sought to use the series to teach the EPO's personnel, especially the examiners, "that patents were no longer limited to technical and legal subject matter. [They] should be looked at in broader context."[109] This training effort also emphasized the relevance of social and ethical expertise to improved work output, although it was not explicitly connected to the patent decision-making process. In addition, Pompidou initiated a scenario-planning project for the EPO, which was devoted in part to assessing the meaning and implications of the growing controversies over intellectual property (which I discuss in detail in chapter 5).

The most formal institutional change, which influenced the examination process directly, was the creation of the Sensitive Cases, or SeCa, system. In its basic approach, it seemed quite similar to the PTO's SAWS program, but its structure and implementation demonstrated how differently the two systems had come to think about morality by the early 2000s. Still in place today, the SeCa system begins at the classification stage. If a patent application is assigned to a technical department seen as particularly controversial (e.g., biotechnology, nanotechnology), as determined by special Issue Management groups made up of EPO employees, the Receiving Section marks it with a small sticker (figure 3.4). The sticker provides a series of prompts meant to guide examiners through an assessment of an application's sensitivity and procedures for additional review, which she conducts in addition to her evaluation of scientific and legal patentability. One of these prompts requires an assessment of whether the application might be in violation of the *ordre public* clause of the EPC and its elaboration in the BPD. Usually, the examiner judges the sensitivity of an application using examples provided by supervisors and her own intuition.[110] The only exception is the weighing-up test that examiners must perform when considering the patentability of animals. One supervisory examiner explained how his subordinates performed this assessment: "It's a bit in the discretion of the examiner to decide what he thinks is [sensitive.] . . . But it's a balancing act between the patient and for the suffering animal. And of course

(I) EXAMINER

1) File contains potentially sensitive matter:

yes ☐ ⟶ SeCa submitted and printed ☐
no ☐

name: .... date: ....

2) Potentially sensitive matter detected after negative finding in 1:

⟶ SeCa submitted and printed ☐

name: .... date: ....

3) Application refused ☐ ⟶ SeCa updated ☐

name: .... date: ....

(II) DIRECTOR/EXPERT (before grant):

File proceeds to grant as proposed by ED ☐

- SeCa Advisory Board consulted ☐
- IMG informed ☐

name: .... date: ....

Figure 3.4.
Sticker affixed to EPO patent applications in controversial areas, as part of its Sensitive Cases system (date unknown).

in some cases, for a cat if you have cosmetic purposes and if that causes suffering [it clearly does not pass the weighing-up test]. But in other cases it might be more difficult for you to decide if it has a small medical benefit and you have a lot of suffering."[111] What is striking about this description is the focus on the examiner's discretion. Unlike the PTO, the EPO seems more comfortable with its reliance on the judgment of examiners, particularly in assessments of sensitivity and *ordre public*. It attempts to inform these judgments through the *Gazette* and seminars, but EPO officials appear to trust the decisions that examiners make. While PTO examiners also make important judgments, the role of discretion in this work is usually dismissed: decisions there are seen as made purely on the basis of science and the law.

If the examiner finds it difficult to make a decision on sensitivity as she fills out the SeCa sticker, especially if it is difficult to determine whether an application violates the *ordre public* clause, she can invite additional review from a member of the patent-law division. The patent lawyer advises her as to whether the application violates the legal precedent established for the morality exception. It is important to note that even when examiners are concerned about the morality of a proposed invention, they can only turn to one of the EPO's established forms of knowledge, in the form of an internal *legal* expert. Given the centrality of scientific and legal expertise in the history of patent systems, and the EPO's interpretation of the *ordre public* clause in the Oncomouse case, this is not surprising. However, this approach limits the degree to which other forms of knowledge or reasoning can figure in patent decision making, even if the bureaucracy has begun to consider ethical reasoning relevant elsewhere.

Supervisors at multiple levels review and certify the application if the examiner ultimately makes a determination of sensitivity.[112] If they agree with the examiner's assessment, it is registered in the SeCa database, which is available to all of the EPO's employees. A specially designated SeCa Advisory Board, made up of examiners, patent lawyers, and public-relations personnel inside the EPO, periodically reviews the database so that it can be aware of the progress of these applications and prepare if the examiner decides to issue a patent based on that application.[113] Meanwhile, the examiner proceeds with the review process. If she made the sensitivity determination because she believes that the application is contrary to the *ordre public* clause, she will reject it and advise the inventor about the problem. The inventor may decide to revise it to avoid the problem, or to challenge the rejection. This process continues until the patent is either issued or rejected. Overall, officials hope that the SeCa system will encourage examiners to scrutinize their work more carefully and ask themselves: "What's happening here? Do we have a quality problem? Are you granting patents you shouldn't be granting?"[114]

Despite the superficial similarities between the SeCa and SAWS systems, they are completely different in practice. SeCa is formalized and systematic. It is embedded in a series of efforts designed to sensitize examiners to the implications of patents and to demonstrate to the public that the EPO understands and is responsive to their concerns. Its focus on controversial areas of technology, as opposed to individual applications, as in the SAWS system, reflects validation of public concern; unlike the US system, it implicitly accepts the view of civil society organizations that certain categories of technology may require different forms of governance. These priorities are also clear in the system's standardization and oversight. The requirements of filling out the SeCa sticker, entering sensitive cases in the database, and the meetings among high-level officials to discuss these cases ensure that the idea of particularly sensitive areas of technology remains alive in the EPO's culture. It may even, as some EPO officials hope, lead rank-and-file examiners to learn more about why some types of patents are so controversial and to consider their decisions accordingly. For their PTO counterparts, however, this would be seen as the worst possible outcome, as it would be understood as bringing subjective judgment to an otherwise objective process. But with these efforts, the EPO redefined its responsibilities to include not just objectivity but explicit consideration of what it sees as the broad implications of patents.

CONCLUSION

In both the United States and Europe, the animal-patent debates had begun with similar concerns regarding the potential for patents to commodify life and validate an area of research that many considered to be unethical. Civil society groups developed similar strategies to transform legal and bureaucratic processes into opportunities to force attention to their concerns. Ultimately, both jurisdictions allowed animal patents, but European patent law articulated some limits.

Looking beyond the question of patentability, however, we see that these controversies shaped who could participate in the patent system and how they could do so, and they continued to influence how the two places understood life, morality, and their governance. The United States articulated a narrower and more technical understanding of patents and their governance than ever before, rejecting wholly the idea that patents themselves had any implications beyond stimulating innovation and markets. This fit with the idea that the government's role was to facilitate the creation of markets by identifying inventions through objective patent decisions. The market would then operate rationally to produce social benefit. To support this, decision makers and traditional interests erected rhetorical, legal, and bureaucratic "expertise barriers" to keep out civil society groups and the alternative forms of knowledge and expertise they brought with them. Only market players and their representatives were deemed relevant and legitimate. And while the PTO developed a new organizational form to respond to public scrutiny, including concerns related to the ethical implications of patents, the domain's techno-legal orientation essentially ensured that the SAWS system would have minimal importance and eventually fail.

When confronted with the moral concerns of patent opponents, the Europeans also drew boundaries around their system. But rather than trying to maintain a techno-legal orientation, as their US counterparts did, they defined moral implications as direct and therefore relevant. Here, government's job was to shape the market to ensure that it reflected the moral standards of society. With this approach came the acceptance of civil society groups and individual citizens as official observers and opponents. The EPO also expanded its definition of relevant knowledge and expertise, as well as its approach to evaluation. It incorporated the weighing-up test and the SeCa system into its examination process and brought in bioethicists, philosophers, and social scientists to speak to EPO personnel. But it had trouble incorporating these alternate forms of knowledge and expertise into decision making, and instead relied on an abstract legal interpretation of the *ordre public* clause to conduct its ethics assessments.

As these debates proceeded, it also became clear that the two jurisdictions approached the governance of life differently. US decision makers continued to refuse the idea that they were dealing with matters of life at all. Even when the PTO rejected Newman and Rifkin's chimera application because it covered a human, it studiously avoided any consideration of how it should value life in its governing practices. To the US patent system, genetically modified animals were technologies, and their status as life forms was immaterial. In Europe, the picture was completely different. Genetically modified animals were understood as having dignity, which needed to be preserved in the process of patenting and commodification. All animals deserved special treatment in the form of a weighing-up test. They were not like any other technology. But the animals that warranted special protection were those whose suffering outweighed the benefits they provided to society. In these cases, the patent system took additional steps to discourage research and markets by prohibiting patents. It is unclear how often the EPO prohibited patents on this basis, although the tests often seemed to justify narrower patents, as they had in the Oncomouse case. Three high-ranking EPO officials referred to at least one example of a patent application that the bureaucracy rejected: a mouse genetically engineered to suffer from baldness.[115]

At this stage in the analysis, we can begin to ask why similar efforts to introduce bioethical concerns into patent decision making resulted in such different legal, technological, and institutional outcomes in the two jurisdictions. I suggest that while both the United States and Europe thought of patents as pivotal to their innovation systems and valued procedural objectivity, their different political ideologies and legacies in terms of how their patent systems envisioned the relationship between government and the market played an important role. In the United States, where markets are seen as the best avenue to producing the public interest, efforts to maximize market efficiency and certainty facilitated strict ideas about legal standing and about who could participate in patent decision making. These formal and informal rules not only made it difficult for nontraditional stakeholders to participate but also provided pro-patent insiders with rhetorical ammunition to suggest that these uninformed and irrational participants were disrupting an otherwise predictable and objective process. It also bolstered their conclusion that patents were amoral objects. Market players would address concerns about human and animal dignity, for example, through the normal operations of the marketplace.

By contrast, there was greater comfort in Europe with the idea that the patent system, guided by an external sense of public morality, should shape the marketplace proactively. The market, in this view, could not always produce decisions that were the best for society and public morality, so government had to step in. So the EPO developed several ways to sensitize bureaucrats and emphasize the importance of bureaucratic judgment on these fraught issues. European governments had done this kind of market-shaping work in the past, even inside the patent system. But the life-form patent controversies established a new role for citizens in the face of growing mobilization related to biotechnology. Decision makers didn't simply interpret the *ordre public* clause and the weighing-up test on their own. Instead, they treated public responsiveness as an important value and invited a broad array of publics to weigh in on decisions that would ultimately shape the marketplace. Indeed, by the end of the animal-patent controversies, the US and European patent systems had begun to define good decision making quite differently.

In reviewing these events, however, it is important to note that these political ideologies did not make the US and European responses to life-form patents completely predictable. Even after the *Chakrabarty* and BPD decisions, the US Congress tried to pass restrictions on animal patents, and the EPO approved the Oncomouse patent. Rather, these ideologies informed and constrained decision making, rendering certain kinds of strategies and decisions more likely and successful. As we will see, the next steps in the life-form patent controversies would not follow a preordained path either. But they would be clearly shaped by how the two places understood the relative roles of government and the market in shaping morality and innovation.

4

FORGING NEW PATENT POLITICS THROUGH THE HUMAN EMBRYONIC STEM CELL DEBATES

By the time scientists began to apply for patents covering human embryonic stem cells in the late 1990s, the US and pan-European patent systems had established different understandings of patents, of appropriate participants, and of acceptable types of participation. They had also interpreted their roles and responsibilities, particularly as they related to the consideration of bioethical issues, quite differently. But it still wasn't obvious what the two patent systems would decide regarding hESCs. Indeed, by then this area of research was provoking furious moral debate in both the United States and Europe.[1] For decades, scientists had explored the possibility that human embryonic cells and tissues, which could grow into numerous types of specialized cells, could regenerate adult human tissues that had been damaged by disease or injury. Now scientists had managed to isolate hESCs, which had the plasticity to turn into any kind of mature human cell, from early embryos. They could potentially produce new treatments for a range of diseases, from Parkinson's disease to diabetes.[2] However, both research into the promise of hESCs, and the development of hESC-based therapies, involved the destruction of human embryos. This had provoked citizen mobilization and government action in both places to protect human dignity and the sanctity of human life.[3]

Human embryonic stem cell patents would ultimately experience different fates in the two jurisdictions: prohibited in Europe, they would be allowed in the United States. But the hESC patent debates were significant far beyond these legal outcomes. We saw in the last chapter that the two patent systems had begun to recognize different kinds of publics and designate different forms of participation as appropriate. But the hESC patent debates would catalyze different political environments for the two patent systems, establishing different kinds of organized interests, different citizen expectations of the patent systems, and different acceptable topics for debate. They would also reflect very different understandings of the role of morality in the patent system, as well as the distribution of moral responsibility in innovation more broadly.

In the United States, new critics—including biologists—emerged after the repeated failures of previous life-form patent challengers. They expressed different concerns from their predecessors. Departing from previous worries regarding commodification, these scientists and their supporters argued that hESC patents might actually stifle innovation, challenging the age-old idea that patents were innovation drivers. By calling into question whether the patent system was consistent with the US's market-making ideology, they managed to penetrate the patent system's barriers and force extensive discussion. Meanwhile, despite the *ordre public* clause and the BPD, Europeans initially permitted hESC patents. This stimulated sizable protests, led by the activists who had been challenging European life-form patents for decades, and resulted in an official EPO opposition. Eventually, the EPO decided that hESCs were unpatentable if they involved embryo destruction, and the European Court of Justice agreed. In this process, European life-form patent challengers became much more systematic in their approach, and established themselves as stable institutional watchdogs who would gain acceptance from European patent system institutions.

HUMAN EMBRYONIC STEM CELL PATENTS
AND THE RIGHT TO RESEARCH

To understand the hESC patent controversy in the United States, we must start with the conflict over patenting complementary DNA (cDNA) that developed during the Human Genome Project (HGP). In the late 1980s, governments across the world established an international initiative to map and sequence the human genome, led in the United States by the National Institutes of Health at a cost of more than $3 billion.[4] Scientists and policymakers argued that it would enhance the world's understanding of human biology and facilitate the diagnosis and eradication of human disease. To do the research, they used a fairly well-known and standardized process to generate DNA fragments that complemented the DNA of the human genome (thus, cDNA).[5] They then used these fragments as probes, also known as expressed sequence tags (ESTs), to figure out the genomic map and sequence.[6]

In October 1991, J. Craig Venter, a molecular biologist at the NIH, with the help of the NIH Office of Technology Transfer, submitted patent applications to the PTO covering thousands of these ESTs.[7] Venter seemed to be operating under the same presumption that had guided passage of the 1980 Bayh-Dole Act: patents were necessary to encourage industry investment in particular areas of innovation, even if they were issued on the fruits of federally funded, early-stage research. As Venter explained later, "When we prepared to first publish our discoveries, we realized that gene sequence disclosure [without patent protection] could threaten development and future protection of new therapeutics because patent law principles preclude protection of obvious discoveries. Our data could render future gene-based discoveries obvious."[8] In other words, Venter sought to patent the cDNA because he believed that it would facilitate future innovation.

Venter and his colleagues likely thought that cDNA was patentable because the PTO had been issuing patents on human genes for almost a decade.[9] Human genes met patentability criteria when the application included a description of the gene sequence and information about the gene's function. Patent applications covering cDNA would look similar in terms of novelty and inventiveness; unlike human genes, however, cDNA probes had no known association to a disease or other conditions. Their "utility," for patenting purposes, came from their use as ESTs.[10]

Leading figures in the scientific community disagreed with Venter's actions. At first, they protested individually, with public statements in the media and in the opinion pages of *Science* and *Nature*.[11] But then they began to organize: professional associations issued statements, scientific leaders testified in congressional hearings, and, eventually, both individual scientists and scientific organizations submitted public comments at the PTO. This was the first time that scientists had mobilized en masse to influence the patent system. While a handful had supported Rifkin's activities, most had seen his and his allies' efforts as radical and fundamental challenges that, if taken seriously, could destabilize their own work.[12] In contrast, these scientists argued that their activism against cDNA patents actually supported the research enterprise.

Scientists focused on two issues. They echoed the first examiners employed by the US Patent Office 150 years earlier, arguing that the cDNA patents involved no ingenuity and therefore did not warrant patents. James Watson, the co-discoverer of DNA, who led the HGP at the time, called the move "sheer lunacy" and argued that "'virtually any monkey' could produce—with the help of DNA sequencing machines—these cDNA probes."[13] They saw patents as similar to scientific publications, which were generated in the wake of an innovative contribution. And in that context, the mere production of cDNA did not merit the reward that a patent represented. To them, ESTs were simply well-developed research tools. If they had understood ESTs as the patent system did, as discrete new objects that were necessary to map and sequence particular human genes, they would likely have had a different opinion of EST patentability.

These scientists also suggested that cDNA patents might have an "inhibitory effect" on research. In other words, these patents threatened the professional and economic interests of the scientists and consequently, they argued, the public interest. The American Society for Human Genetics, for example, explained:

> Normally, a patent ensures that a gene will be available for all researchers and for any company willing to license it. We fear that in the case of ESTs it may have quite the opposite effect. An EST patent, to be useful to the commercial sector, must make broad claims in regard to future use, including protection for the rest of the gene and its protein product, and their use for diagnostic and therapeutic applications. The academic community is unlikely to put major research effort into an EST-identified gene or its protein product if someone else already has the right to license its use based on the trivial effort required to sequence the original EST. In the commercial sector there may be reluctance to invest heavily in further research on EST-identified genes when a small but unknown fraction of them will turn out to have commercial utility, and when the useful ones may be contested by patents involving other ESTs from the same gene. Genome research could end at the level of ESTs."[14]

This argument would be powerful, particularly in comparison to those made by the previous generation of challengers in the US patent system. Like Rifkin and his allies, these scientists were challenging a basic assumption of the patent system—but they focused on the idea that patents stimulated innovation. And they based this argument on their own social and economic understanding of their professional fields. Unlike Rifkin, however, they possessed formal, technical training that was valued in the patent system and highly respected in the US context more generally. This gave them more credibility.[15] They also had clear and direct economic interests that might have made them more comprehensible in a political environment that was most receptive to the interests of market players. But perhaps most importantly, they were framing their concerns in terms that would connect to deeply held American ideals about promoting innovation and scientific progress.[16] If these patents really hurt research, this would strike at the heart of the US approach to innovation policy.

Indeed, policymakers began to pay attention. In 1992, Congress held hearings entitled "The Genome Project: The Ethical Issues of Gene Patenting." Given this title and the previous life-form patent battles in the United States, we might have expected testimony focused on the ethical implications of turning cDNA into a patented commodity. But these hearings marked a slow shift away from these topics. Except for Andrew Kimbrell of ICTA, who had been active in the earlier animal-patent debates, the witnesses seemed more like traditional patent-system participants. They included representatives from patent-law and biotechnology companies. Venter and the NIH director also testified. For the most part, these witnesses focused on the concern that cDNA patents would stifle research. In identifying the main issues, Bernadine Healy, director of the National Institutes of Health, raised the deontological questions that had preoccupied the previous generation of activists but spent most of her time on the new issues regarding innovation: "Should we patent what has been called our universal heritage? Should the US Government, rather than individual scientists, be filing or holding patents on cDNAs? Should the NIH exert its authority over grantee institutions on DNA patent filings? Would patenting and licensing of gene sequences help or hinder product development?"[17] The rest of the hearing also reflected this shift. Although Kimbrell and the senators occasionally raised concerns about taking "the common heritage of all of us" and allowing it "to be corporately enclosed,"[18] and about "deep-rooted religious and ethical questions [that] surround the patentability of life,"[19] those arguments generated little discussion. Rather, the hearing focused on Venter's patent applications and whether they, and DNA patents more generally, might stifle innovation because they were being issued at early stages of research.[20]

The cDNA patent debate proceeded for years and continued to focus on research impact. Concerns about the commodification of life forms through patents seemed secondary at best, and the civil society groups who had raised these issues seemed mostly to have moved on to other topics. In 1999, the PTO proposed revised guidelines that validated the patentability of cDNA sequences in theory but emphasized the importance of the "utility" criterion, requiring each application to demonstrate that an invention was "specific," "credible," and "substantial." Until then, utility had not been as serious a consideration in the examination process as novelty, inventiveness, or the description in the application, probably because the latter criteria were easier for technically trained examiners to judge.[21] The PTO proposal generated public comments from the customary participants in the patent system and from some of the advocacy groups that had led the first round of challenges on life-form patent debates, but also from its newly self-identified stakeholders: scientific associations and individual scientists.[22]

The PTO summarized them: "Several comments state that . . . patents on genes are delaying medical research and thus there is no societal benefit associated with gene patents. Others state that granting patents on genes at any stage of research deprives others of incentives and the ability to continue exploratory research and development. Some comment that patentees will deny access to genes and our property (our genes) will be owned by others."[23] The PTO rejected all of these concerns, and focused its response on the charge that patents might stifle innovation. Much as it had responded to earlier charges regarding the ethical implications of commodifying life through patents, the PTO argued that the commenters simply didn't understand how the patent system worked. It noted, "The incentive to make discoveries and inventions is generally spurred, not inhibited, by patents. The disclosure of genetic inventions provides new opportunities for further development. . . . Other researchers may discover higher, better, or more practical uses, but they are advantaged by the starting point that the original disclosure provides."[24] To the PTO at the time, it was inconceivable that patents themselves might hurt innovation, or that patent holders might act in ways that hurt research overall.

The PTO formalized these guidelines in 2001, and the NIH withdrew its applications for the cDNA patents, but the episode had a political impact beyond cDNA patents and even beyond human genetics research. It had generated a new policy concern focused on whether life-form patents could hurt biomedical research (and innovators would eventually raise similar concerns regarding other areas of technology as well).[25] It had also reframed the life-form patent debate to focus on the pace of innovation rather than moral and socioeconomic matters. This framing may have seemed more politically palatable because it focused on the concerns of scientists, who had cultural credibility, and their desire to stimulate innovation, which was central to the US imagination. But, it also challenged the harmonious picture of innovation that decision makers and the patent system's organized interests held dear.

Producing New Knowledge, New Experts, and a New Science Policy Debate
The issues that emerged in the cDNA debate were not, however, completely new. Scholars had raised concerns that patents might hurt academic biomedical research since the late 1980s.[26] They worried that too many patents, issued early in the innovation process, could become hurdles if patent holders wielded their rights too aggressively or scientists had to negotiate multiple patent licenses before conducting their research. The cDNA controversy brought new attention to these arguments and stimulated the development of an academic subfield.

One of the earliest scholarly critics was Rebecca Eisenberg, a law professor at the University of Michigan. Identifying the problem in a series of papers, she recommended that it be addressed by reviving and clarifying the "experimental use" exemption that had first appeared in a Supreme Court decision in 1913.[27] The justices had then opined that "it could never have been the intention of the legislature to punish a man, who constructed a [patented] machine merely for philosophical experiments, or for the purpose of ascertaining the sufficiency of the machine to produce its described effects."[28] But while the courts had recognized the exception in theory, defenders in infringement lawsuits had almost never succeeded in practice.[29] So Eisenberg articulated specific circumstances in which an experimental use exemption should be allowed. She published this work primarily in legal journals, which initially garnered little attention from scientists or policymakers. But as the cDNA patent controversy developed, they began to notice. In 1998, Eisenberg and a coauthor, Michael Heller, published a piece in *Science* that used the cDNA patent case and others to explain how patents could hurt innovation in biotechnology. Entitled "Can Patents Deter Innovation? The Anticommons in Biomedical Research," the piece was widely read and frequently cited.[30]

Eisenberg's work also inspired others to investigate the issue. Soon, other legal scholars began to publish analyses of how patent law might be reshaped if patents did, indeed, hurt innovation.[31] And bioethicists and social scientists generated empirical data specifically related to patents in biomedicine and biotechnology. Some conducted case studies of the impacts of patents on human genes, finding that, under some circumstances, patents might, indeed, dissuade researchers from working in a particular field or on particular scientific problems.[32] Others organized surveys, with one concluding that according to geneticists, patents inhibited their research.[33] Economists and innovation-policy scholars initiated quantitative studies that produced mixed results: some validated the case study and survey research, but some showed no effect, or at least a murkier picture.[34]

The National Research Council of the National Academy of Sciences Complex, the central external advisory body for science and technology policymaking in the United States, responded to these growing concerns as well.[35] Starting in the early 1990s, its expert committees—which often included Eisenberg—published multiple reports on the subject.[36] What was clear by this point, particularly in comparison to the earlier generation of debates, was that questions regarding commodification, which challenged the morality of both innovators and the innovation process, had fallen largely away. The new scholarly and policy debate focused on the impacts of patents on research. This framing excluded the experts and participants that Rifkin and his allies had tried to bring in, and it maintained innovation's morally privileged place.

Whichever side of the new debate they were on, participants saw the stimulation of innovation in biotechnology as a good thing.

As we will see later in this chapter, questions about the impact of patents on research did not stimulate the same kind of scholarly and policy debate in Europe, partly because Europe had no analog to the Bayh-Dole Act, but also because it had addressed many of these issues years before. European countries had long legal histories of an "experimental use exemption" in their patent laws.[37] By the late nineteenth century, the UK courts had decided that an experiment without direct commercial intent, which was designed to improve "the invention [that is] the subject of the patent, or with the view of seeing whether an improvement can be made or not . . . is not an invasion of the exclusive rights granted by the patent."[38] By the middle of the twentieth century, other European countries had established similar approaches; for example, Germany's 1968 Patent Act permitted experimentation on patented inventions so long as there was no "further commercial motivation."[39] In 1975, an experimental use exemption became part of European Community law. In the years afterward, it was interpreted broadly in both case law and national statutes.[40] In Europe, any risk that patents could hurt research had been deemed too great and warranted preemptive action. This regulatory strategy echoed Europe's approach in many other domains, but it also rested on a fundamental difference between the two systems that I have noted often above: in the European context, patents were seen as having both positive and negative implications, and it was the government's role to act to maximize their benefits.

The hESC Patent Question Emerges

As the academic and policy debate over the impacts of patents on innovation was flourishing in the United States, hESC research was developing. In 1998, James Thomson of the University of Wisconsin announced that he had isolated and maintained hESCs in culture, which was seen as an important breakthrough.[41] On his behalf, the Wisconsin Alumni Research Foundation (the patenting and licensing arm of the University of Wisconsin) applied for patents on this work, including the hESCs themselves. Once they were granted, WARF licensed them exclusively to the biotechnology company Geron.[42] Geron, in turn, established expensive licensing terms for users—including university researchers. For scientists and others already concerned about the impacts of patents on biomedical research, hESCs provided a perfect example. It was a relatively new field, and Thomson had applied for patents covering foundational innovations that could limit research dramatically if they were wielded aggressively. For researchers already worried that the controversy over the moral status of hESCs would restrict their work, patents created another access problem.

The Thomson/Geron patents, and hESC patents generally, stimulated a great deal of discussion focused on the potential negative implications for research. But what is especially curious is that the issues that were capturing US policy and public discussion regarding hESC research, which focused on the moral status of the embryo and the ethics of destroying embryos to produce medical treatments, emerged rarely in the conversations over patents. There seemed to be implicit agreement, particularly after the early life-form patent battles, that the patent system was not an appropriate place for governing these "ethical" issues.

Consider the Senate hearings that took place soon after Thomson announced his findings. Arlen Specter, chair of the Senate Subcommittee on Labor, Health and Human Services, Education, and Related Agencies, convened three days of hearings to discuss the governance of hESC research and of related technological developments. A Republican from Pennsylvania, Specter was a strong supporter of biomedical research and NIH funding, and was well aware of the ethical concerns regarding hESC research. In his opening statement in December 1998, he made a significant distinction: he separated "ethical" concerns regarding hESC research from those focused on whether the hESC patents would stifle innovation. He noted: "The discussion which we will be initiating today, or carrying forward today, is one which will challenge ethicists and theologians as well as Senators and members of the House. The *collateral* question arises as to whether these procedures may be patented [emphasis added]."[43] This was quite a different approach from that of the congressmen who had initiated hearings regarding animal patents in the 1980s, who had structured the schedule to consider the ethical implications of patenting genetically engineered animals.

To speak to the intellectual property issues, the subcommittee invited the director of the Office of Technology Transfer for the National Institutes of Health, the acting commissioner of the Patent and Trademark Office, a representative from the Biotechnology Industry Organization, and stem cell scientists. Senators framed this portion of the hearing by echoing the scientists' concerns that hESC patents might hinder research. Democratic Senator Tom Harkin of Iowa stated, for example, "The dilemma is, we want the research, we want the money invested, we want private moneys invested, and we also want to make sure that they are able to get a return on that investment and that they are able to patent it, but we also want to make sure the research is broadly and widely available to others, and that we are not hindered by a broad patent that is issued that hinders further research that may be utilized."[44]

Harkin expressed the concern that originated in the cDNA controversy, that patents and innovation could be at cross-purposes. But the witnesses, many of whom had appeared in earlier congressional hearings regarding life-form patents, responded by arguing that patents were quite limited. As his predecessors had done before, the acting commissioner of the PTO argued, "It is not a monopoly right to own an invention as is sometimes suggested, but rather it is the right to exclude others from making, using, or selling it, and at the patent owner's discretion that right may or may not be exercised."[45] This testimony reinforced the idea that patents were merely techno-legal objects that conferred limited rights to the owner. They did not confer a positive ownership right, and therefore could not, by themselves, have social implications. It also reiterated the view that those who claimed that patents hurt research did not understand how the system worked. While some witnesses did validate the senators' concern, they argued that it was relatively minor and would be solved by the rational choices of market players. The NIH representative noted, for example, "It is our view that these licenses can be crafted to ensure [that] commercial and research purposes be both preserved."[46] In this view, patent holders were likely to operate in the public interest, licensing the patent widely, because it would be in their own best interest. It would ensure that they would reap more revenue.

Congress took no action at this time, and in August 2001, President George W. Bush issued an executive order requiring any researcher working with federal government funds to use hESC lines that had already been produced.[47] This had the potential to dramatically limit federally funded hESC research. Citing similar moral concerns, some US states began to adopt restrictive laws that simply banned the creation and destruction of human embryos for research purposes.[48] Meanwhile, Geron imposed its high licensing fees on any scientists conducting hESC research.[49]

Together, these events increased scientists' worries that the pace and scope of hESC research were suffering. Some even began to speak in terms of "freedom" and "rights" to research, suggesting that they were akin to Constitutional free speech rights.[50] The same Senate subcommittee, now run by Senator Harkin, held hearings on stem cells again in 2001. This time, many more witnesses articulated concerns about the impact of patents on research. University of Pennsylvania bioethicist Glenn McGee worried that "if stem cell research is tied up at this stage by patents and licensing agreements, even if those patents are held by universities who trade them in fair and honest ways, the effect will be to hamper and slow research, but moreover, it will tax any Federal dollars for stem cell research in what I think you could argue is an unacceptable way."[51] This statement, coming from a bioethicist, is notable. It demonstrated how the patent debate—and even the concerns of bioethics—in the United States had shifted away from arguments related to the commodification of life forms through patents to a discussion about how patents might be immoral because they hurt research. After all, bioethicists in the earlier animal-patent hearings had focused on concerns regarding animal welfare and on the idea that patents implicitly validated controversial research. The statement also shows that, within the patent system, hESCs were defined as a fundamentally moral and socially beneficial innovation. The challenge, as the bioethicists saw it, was to produce as much hESC research as possible.

The traditional supporters of the patent system responded as they had in the 1998 hearing, suggesting a very limited definition of patents. They argued again that if any access problems did arise, they were the responsibility of the market and would be quickly resolved by individual market players.[52] As though it was validating this ideology, two months later WARF announced that it would allow all NIH grantees to license its hESC patents for a relatively low one-time fee.[53]

Even as these congressional hearings demonstrated a significant shift in discussions over life-form patents, commodification concerns had not yet disappeared completely. Some of the congressmen who were the most outspoken on abortion tried on a couple of occasions to restrict the patentability of hESCs and human clones, but their efforts were limited and sporadic, and ultimately unsuccessful. In 2002, Republican Senator Sam Brownback from Kansas suggested an amendment that would ban patents on the processes and products of human cloning (which could include hESCs). Placing his concerns in the context of the ethical debate over human cloning, Brownback explained, "Shall we use human life for research purposes? Shall we use human life for commercial purposes? . . . In this debate we will have to answer whether or not the young human at his or her earliest moments of life is a person or is a piece of property."[54] His amendment never came to a vote, and Brownback did not try again. Two years later, another abortion opponent, Representative Dave Weldon, a Republican from Florida, achieved more success when he amended an appropriations bill prohibiting the patenting of humans. But when the Biotechnology Industry Organization and the PTO opposed it,[55] Weldon backed down and stated that his intent was not to change existing patent policy. So although the amendment passed, the PTO commissioner noted that he understood the Weldon amendment to be identical to current US policy.[56] Although the moral dimensions of life-form patentability concerned these congressmen, even they seemed reluctant to discuss concerns related to human dignity in the context of the patent system.

In this discussion, just as in the case of the Newman-Chimera patent application, the PTO had refused to interrogate its definition of a "human being." We can imagine a couple of reasons for this. Initiating a lengthy discussion about the ontology of a human being, particularly in the midst of the debate over hESC research, would have validated the idea that the patent system was engaged in the ethically fraught task of governing life. This would contradict the patent system's previous efforts to focus the discussion on whether life forms constituted nature or technology, which had allowed it to focus on scientific evidence and legal reasoning. Furthermore, by this time many scientists had received patents related to human cloning and hESC research.[57] Questioning how the system defined a human being would have jeopardized these patents and therefore the assumptions and expectations of market players.

Ultimately, these and other senators focused their concerns regarding the sanctity of human life on trying to ban hESC research or at least to restrict funding dramatically.[58] And while the anti-abortion movement occasionally expressed concern about hESC and cloning patents,[59] they would not make these issues a priority either. They too focused on research policy. These political calculations reproduced the idea that the patent system was a technical domain focused on stimulating innovation and markets, designed to facilitate the needs of the patent holder. These ethical concerns were to be addressed elsewhere.

The Debate Moves to California

The battle between researchers and patent holders was far from over. By the mid-2000s, WARF and Geron had made it easier for scientists to license their patents, and the initial concerns had eased. But many were still worried about their freedom to do research, which, they argued, both the federal research restrictions and patents could jeopardize. With this in mind, a few states developed hESC research-funding programs of their own. California's was the largest and most ambitious.[60] In 2004, its voters approved Proposition 71, a referendum authorizing the state to spend $3 billion in bonds to fund hESC research. The referendum also established a "right" to conduct stem cell research,[61] and it characterized intellectual property as potentially infringing upon this right. In doing this, it formally introduced a skepticism regarding the patent system that was rare in US policy. It advocated a "balance" between providing researchers and patients access to research results, on the one hand, and economic benefits to innovators and the state of California, on the other.[62] It then took on the responsibility of assuring "that essential medical research is not unreasonably hindered by the intellectual property agreements."[63]

This, in and of itself, was an important turn of events. In the United States, enough people, and particularly people with cultural legitimacy, had begun to question the relationship between patents and innovation that it had become an area of policy concern for the country's most populous state. This was a significant shift for California, which had been a strong proponent of patents in the late twentieth-century congressional debates over the Bayh-Dole Act due to its purported benefits for the research endeavor.[64] Now, however, California itself seemed to define patents as having both positive and negative effects on innovation and health care. Patents could hurt the public, and the state seemed to understand its role as minimizing these problems while also maximizing the benefits for patent holders.

As soon as the stem cell referendum passed, the state established a Task Force to develop an intellectual property policy for the new California Institute for Regenerative Medicine (CIRM), which would dispense the research funds; clearly, patents were a priority. The CIRM chose members from the university, industry, and patient advocacy communities—all of whom had become important stakeholders in US discussions related to biomedical policy—to sit on the Task Force. Together, they were supposed to codify the referendum's balancing approach in a policy that would govern all of its grants, contracts, and loans to both nonprofit and for-profit organizations. And their mission had just become more pressing: in 2004, the US Court of Appeals for the Federal Circuit had ruled that, given current patent laws and case precedents, a researcher could only enjoy an exception to patent law on the unlikely occasion when her work was out of "idle curiosity."[65] Realistically, this meant that all researchers who used patented inventions had to obtain a license or risk patent infringement.

At the first meeting, Task Force members expressed concerns not only about the potential negative impact of patents on innovation, but also about their potential impacts on health care access. Dr. Jeannie Fontana, a physician and biochemist who served as the executive director of patient advocacy at the nonprofit Burnham Institute for Medical Research, suggested a collaborative approach to research: "Perhaps CIRM could maybe come up with a new model where we could take advantage of more collaborative efforts, I know it's idealistic, in how to deal with all the property issues, but really where we incentivize collaborations is more heads together are better."[66] Jeff Sheehy, a longtime AIDS activist who worked as the director of communications for AIDS research at University of California San Francisco, emphasized the development of policies that would support widespread access to the resulting technologies: "We're missing in this whole scheme something that really talks to someone who's a patient in California and says that they're going to benefit for foregoing $3 billion that could go into Medi-Cal tomorrow, they could go into healthy families tomorrow, and that's what we're missing in this equation."[67]

Not surprisingly, California's biotechnology industry and patent-law community immediately challenged the CIRM's efforts to develop a new approach. Like many of the witnesses who had testified in the earlier Senate hearings, they argued that, overall, the existing patent system worked, by which they implied that it generated socially beneficial innovation. Evidence to the contrary, they argued, was merely occasional anecdotes. The California Council on Science and Technology (CCST), a nonpartisan, nonprofit organization established by the California legislature in 1988, issued a report that sided largely with these stakeholders.[68] Written by a working group that was made up of patent lawyers, an economist, and high-level representatives from technology transfer offices, industry, university, and government laboratories, it concluded that the CIRM should maintain the traditional approach to intellectual property.

Two members of the CCST's working group discussed the report with the Task Force. Steven Rockwood, executive vice president of Science Applications International Corporation, a Fortune 500 science and technology applications company, emphasized the need for patents to stimulate rapid innovation: "The major objective should be to incentivize the adoption of whatever inventions come and get that into the public domain as fast as possible so that these drugs and treatments are available to market. . . . The private concern will put in 90 percent plus of the money that it took to get that drug to market, and there must be some return there."[69] He and other members of the CCST committee tried to dissuade the Task Force from limiting patent rights too severely, while challenging the idea that research could be stifled by patents. In his view, the patent system was part of a linear approach to innovation, in which private interests required patents and licenses to create the incentives that would eventually bring innovations to market.

The CCST members also suggested that the Bayh-Dole system, in which the funder allowed the grantee to own the patents, largely worked. Fred Dorey, a patent lawyer specializing in the life sciences, noted, "So I think the important thing is to understand that if there is a system out there that's working, Bayh-Dole may not be perfect, and there's ways to improve on it, but to try and turn the corner and create an entirely different system in a very new, early stage technology is only going to add traumatic and dramatic complexity and inefficiency and cost to this process."[70] Cases where patents stifled research, they argued, were extremely rare and could be addressed by establishing contracts with grantees that established a right to "march-in" if the patents were not properly licensed (the US government has this right, but has never exercised it).[71]

Both Rockwood and Dorey agreed that the CIRM should fulfill its responsibility to the public interest by ensuring the rapid development of technology, and that this would be best accomplished by maintaining the current approach to intellectual property policy that was based on decades of experience with innovation and the patent system. Just as insiders had rejected previous challengers on the basis of their perceived ignorance, these stakeholders emphasized the Task Force's limited knowledge of the patent system. But they also dismissed Task Force concerns in other ways. They argued that any problems with the system were rare, advocating a utilitarian calculus. They also appealed to the Task Force's sense that patent holders were morally responsible actors who would behave in the public interest (In addition, they argued, it was in the inventor's interest to license the product widely to reap revenues.) Any change to this system would introduce uncertainties that would hurt, unfairly, the innovation community.

Despite this testimony, the Task Force maintained its commitment to a novel approach in its first few meetings. In November 2005, it heard testimony from law professor Rebecca Eisenberg. She advocated a research exemption that would permit researchers funded by the CIRM to use materials patented by CIRM grantees in their work without a license, so long as they did not receive a direct financial benefit.[72] In support of this position, she cited cases of "overly aggressive patenting and licensing strategies for upstream research discoveries of the sort that could otherwise be readily disseminated in the public domain without the need for patents. In some cases the patenting of these discoveries has led to wasteful transaction costs and obstacles to research, perhaps to the long-term detriment of progress in research and product development."[73] Eisenberg clearly did not agree that patent holders always operated in the public interest. Nor did she accept the claim that patents only caused rare problems for researchers. Instead, she emphasized the benefits of hESC research and the rights of scientists to work freely, and thus advocated an experimental use exemption.

Groups who claimed to represent the public interest, including the Center for Genetics and Society (CGS) and the Foundation for Taxpayer and Consumer

Rights (FTCR), agreed with Eisenberg. In several meetings, their representatives gave public comments that encouraged an intellectual property policy that ensured benefits to California taxpayers through efficient management of and access to the technologies that resulted from the research. Although they did not explicitly endorse a research exemption, they sought to ensure that both scientists and taxpayers had widespread access to the findings of and technologies resulting from the CIRM researchers. Jerry Flanagan, from the FTCR, noted, "California has the opportunity not only to adopt an IP policy that protects the intent of Prop 71, but also become a model for national intellectual property research and division of royalties and IP ownership. That's the kind of thing we heard from California is [that it] has the glow of biotech success and [is] the envy of the world. . . . However, from the patient advocacy perspective, the way that California would then move to the pinnacle, in the taxpayer's perspective, . . . would be to devise a policy that not only gets those people to play, but also in some real way provides public benefit and a mechanism for public control."[74] They also cited other cases, particularly from the pharmaceutical industry, to suggest that patents invariably limited access and to advocate a precautionary approach in favor of less patent protection.[75]

None of the public comments raised moral concerns about the creation of hESC or embryo markets due to patents. This was particularly surprising in the case of the CGS, which was a pro-choice, liberal group that had ties to both Rifkin and Kimbrell. It had challenged hESC research in other venues because of concerns that it would commodify human eggs.[76] It may have made this strategic choice because the CIRM and the Task Force were focused on licensing rather than patentability. Given the shifting dynamics of the US patent controversy, however, it is also possible that these groups had concluded that they would be more successful in achieving a policy they preferred if they allied themselves with scientists' concerns.

In February 2006, the Task Force issued a draft policy that embodied its balancing approach and sought to ensure access to researchers and patients. It went beyond the inclusion of march-in rights on individual contracts and included several novel elements: it assured that California research institutions could use patented inventions at no cost; it encouraged broad licensing of patented inventions; it required that inventors share with the CIRM 25 percent of patent-related revenues (minus the percentage paid to the inventor); and it mandated pricing of therapies and diagnostics at or below the federal Medicaid price.

California's biotechnology industry was strongly opposed to this draft policy. It charged that the Task Force lacked the requisite experiential expertise to create it, that the policy would introduce intolerable uncertainties into a system that largely worked, and that the Task Force was creating a false dichotomy between the interests of the biotechnology industry and those of researchers and taxpayers.[77] As it made these arguments, it emphasized its own expertise and the CIRM's lack of knowledge on the topic. It also referred to its own "collective experience" with "hundreds" of cases, comparing it implicitly to the handful of problems cited by Eisenberg and others. The California Healthcare Institute, which represents California's biomedical research industry, noted, "During the past thirty years, California biotechnology companies have licensed hundreds of inventions from academic institutions. The lesson from this collective experience is that stakeholders—researchers and research organizations, industry and other licensees, and venture capital investors—value transparency in licensing and technology transfer agreements. Biotechnology is inherently risky. Any aspect of a technology transfer contract that increases risk, particularly by adding an element of uncertainty, makes it less attractive to potential partners and investors and thus reduces the prospects for successful commercial collaboration."[78]

These opponents did not simply fight one set of cases with another. They suggested a different approach to evaluating these cases. Proponents of a novel policy had argued that even a few examples of stifled research or inordinately expensive therapies—particularly in the early stages of a field—were too many (in fact, the Europeans had established an experimental use exemption decades earlier using this evaluation approach). But opponents in California suggested that the burden of proof should be on the Task Force and others who sought to change the status quo because of the dangers of market uncertainty.

In response, Task Force members softened their positions almost immediately. Jeff Sheehy, who had advocated broad and creative thinking initially, agreed that the Task Force lacked the relevant expertise to propose what he now saw as a risky approach: "This seems to be occupying a lot of really smart people in Washington and other places, and they haven't got the answer yet. And I don't think we—there are places we can exert leadership and ought to. It was bold of us to try in this field, but I'm not convinced, if there's an economic cost to the state of California, that I can in good conscience go forward with that."[79] Others joined him, worrying that the novel approach to patents would hurt the innovation process. Duane Roth, chief executive officer of CONNECT, a company that fosters new business development in the life sciences, suggested that "before we start playing with this, because it's extremely complicated in terms of the unintended consequences and how something like this is interpreted, that we stay the course right now. . . . I'm always reluctant, given our thin staff and amount of review we've had on this, to be out in front on something like this when we've got something that's working extremely well for twenty years."[80]

Eventually, the Task Force eliminated the research exemption from its policies, and softened the other new elements significantly.[81] The insiders had convinced Task Force members that they had insufficient knowledge about the patent policy domain, that the proposed changes could cause major harm to the innovation process, and that their responsibility to the public interest should be conceived solely in terms of the speed of technology development and maintenance of a strong patent regime. And as they eschewed the idea that patents might hurt research, they maintained the moral authority of all researchers, including patent holders.

Rewarding the True Inventor

Dissatisfied by the responses from both the federal government and California, those concerned about research impacts tried to find another venue to challenge hESC patents. The FTCR and Jeanne Loring, a stem cell biologist based at the University of California–San Diego, who had long questioned whether Thomson deserved credit for isolating and propagating hESCs,[82] solicited the help of Daniel Ravicher, who had recently founded the Public Patent Foundation (PubPat). Ravicher created PubPat to "represent the public's interest in the patent system," and he had the appropriate background and expertise to do so.[83] He had an undergraduate degree in mechanical engineering and a law degree, and had spent time specializing in patent law at a private firm. But he had grown disillusioned and frustrated that "the patent system had been skewed to benefit private interests. The public needed a champion."[84]

Founded in 2003 and funded by small donors, private foundations, and angel investors interested in social entrepreneurship, PubPat's staff was extremely small. Ravicher worked with two other lawyers and gained additional help from law students (he holds a faculty position at Yeshiva University's Cardozo Law School). His patent-law expertise made him aware of potentially powerful places to intervene and also provided him with the knowledge and rhetorical resources to make arguments that system insiders would be forced to consider. This expertise, coupled with his commitment to change in the patent system, distinguished him from the earlier challengers and made him a potentially formidable ally for those who sought to assert public-interest critiques of the closed patent system.

Eventually, PubPat would submit public comments on proposed rule changes at the PTO, testify in front of Congress and the European Parliament, submit amicus briefs to the Court of Appeals for the Federal Circuit and Supreme Court, and challenge individual patents at the PTO and in the courts.[85] But the hESC case was one of Ravicher's first, and he decided to file a reexamination request challenging the WARF/Thomson patents, as the ICTA had done with animal patents. And just like the previous patent challengers, he coupled his public-interest concerns with arguments that would be more acceptable in light of the bureaucratic rules. To the PTO, the only relevant question in the reexamination process was whether there was a technical error, based on previously undiscovered "prior art" (usually scientific publications or patents) that called patentability into question. This reinforced both the PTO's techno-legal approach and its procedural objectivity. So, Ravicher coupled his public interest concerns with prior art issues. In his formal requests to the PTO, he described his public concerns as follows:

> These three patents, which broadly claim any primate or human ES [embryonic stem] cell, are being widely and aggressively asserted by their owner against every human ES cell researcher in the United States. . . . This not only harms scientific advance here in the United States, it also has a harmful economic impact on Americans by diverting taxpayer dollars meant for research to pay for licensing fees. . . . Although these scientific and economic concerns are admittedly not grounds to grant this request for reexamination, [the] FTCR respectfully requests that they be considered.[86]

This was, of course, the guiding concern of both Loring and the FTCR. And perhaps they, along with Ravicher, thought that reminding the PTO of these public concerns might shape its decision making, particularly given the attention paid by both Congress and the CIRM to these issues. Under similar circumstances, after all, the EPO had opened up its decision making regarding animal patents. In addition, the FTCR and Ravicher knew that these arguments would likely resonate in the public arena (more than the technical arguments focused on novelty or previous life-form patent challengers' concerns regarding commodification); they emphasized these in public statements on the case.[87]

But the PTO dismissed the innovation concerns immediately: "The third party discussion of Harm caused . . . is clearly outside the scope of reexamination and thus has no bearing on the raising of [significant new questions]."[88] The PTO did, however, grant reexamination on the basis of a "substantial new question of patentability" raised by prior art. In March 2007, the PTO revoked all of the patents. Eventually, these revocations were appealed and the patents were partially reinstated.[89]

This reexamination episode was important beyond these decisions. It emboldened Ravicher and gave him a good reputation among the growing group of scientists concerned about the impact of patents on their research. As we will see in the next chapter, it would increase his momentum and contacts, and allow him to carve out a space for a new kind of expertise, in public-interest patent law. But, it also reinforced the PTO's narrow approach to stimulating innovation, even in the face of growing concern about the relationship between patents and research.

The US debate over hESC patents would not inspire major policy change at the national or state level. But it demonstrated how the system's institutional framework, as well as the shared understandings of its decision makers and regular participants, shaped and constrained the policy discussion that took place. Concerns regarding commodification and attempts to use the patent system to govern an ethically controversial area of research slowly disappeared during this period. Despite the acrimonious battles raging over the ethics of hESC research and treatment, the patent system was never treated as an appropriate or relevant site for this kind of governance. Even interests who had previously raised concerns regarding the morality of patents either kept quiet or shifted their arguments. Debates about the ethics of science and technology seemed to belong in research funding agencies, at the FDA, inside legislatures, and in the public domain. The patent system's only focus was to stimulate innovation, and thus it was only brought into the debate when scientists and others worried that patents might hurt research. But rhetorical and bureaucratic barriers made it difficult to initiate policy change on these issues too.

The hESC patent debates had also introduced a new kind of moral conflict. Historically, both scientists and patent holders had enjoyed similar social and political credibility in the United States because they pursued innovation. Now, however, as their interests had begun to clash, their relative authority became unclear. On the one hand, scientists were becoming increasingly skeptical of patents, which they worried were creating insurmountable obstacles for research and could have disastrous social impacts. Patent holders, on the other hand, argued that innovation would stall if patents were not available as incentives. Furthermore, to not patent could prevent the distribution of socially important technologies. For now, policymakers responded by maintaining their trust in the rationality of the marketplace and in the moral authority of patent holders. In the aggregate, they continued to believe, both innovation and society would ultimately benefit from a strong patent system. This picture, of course, conflicted with the one emerging in the research-funding domain, which suggested that hESC researchers needed clear rules in order to behave responsibly. But the power of the US patent system's vision came from the distance it had managed to maintain from the other domains that govern science and technology.

REGULATING HESCS THROUGH THE EUROPEAN PATENT SYSTEM

Initially, the European Union did not have a uniform policy guiding hESC research; most European countries established their own approaches. In Austria, for one example, all human embryo research, including the derivation of stem cells, was banned.[90] The United Kingdom, for another, allowed both embryo and hESC research, but its Human Fertilization and Embryology Authority (HFEA) oversaw the research according to a strict set of rules.[91] As we will see, however, the patent system soon became a central site for shaping the ethical dimensions of hESC research and technology at the European level.

The European hESC patent battles began in 2000, when the EPO, despite its supposedly growing sensitivity to the ethical dimensions of patents, awarded to the University of Edinburgh a patent covering a method of isolating and propagating animal (including human) embryonic stem cells. Known as the "Edinburgh patent," it also claimed methods for making transgenic animals, including humans. By then, the EPO had been struggling for years to address what it, the European Parliament, and European citizens had identified as the ethical implications of patents. The Oncomouse patent opposition had forced it to contend with the meaning of the *ordre public* clause and with new interests and experts who claimed that they had an important contribution to make regarding the role of morality in the patent system. The EPO had responded by accepting these participants and using abstract legal reasoning to define and apply the *ordre public* clause. But civil society groups continued to pressure the EPO to further define what they saw as the moral, socioeconomic, and environmental implications of patents, and to articulate a standard evaluation scheme. Their primary strategic avenue was the EPO's opposition mechanism, which they used to challenge a variety of patents that they argued were detrimental to public order and morality.

Christoph Then, who would lead the Edinburgh opposition, organized many of these oppositions and had become the primary civil society–based patent critic in Europe. Trained in Germany as a veterinarian, Then had ethical concerns regarding life-form patents that echoed Rifkin's. He worried that these patents would change how humans understood and valued life. Life "was not invented by mankind," he observed.[92] He first engaged with life-form patent issues through the Oncomouse opposition, which he guided on behalf of Greenpeace, and activism against the BPD. In 1992, he created Kein Patent Auf Leben (KPAL; No Patents on Life) to focus on patent challenges.

KPAL achieved mixed results in terms of patent outcomes, but throughout the 1990s, it gained expertise in the patent system both in terms of the specifics of European patent law and in terms of the formal and informal rules of engagement with the European Patent Office. This expertise made it a resource for others interested in challenging patents; KPAL occasionally helped governments and civil society groups in the developing world successfully oppose patents covering medicinal herbs and plants and traditional methods of breeding corn.[93]

This growing expertise helped the organization become systematic in its activism and clearer regarding its goals. KPAL's early choices of which patents to challenge had been ad hoc: it worked against the BPD at the European Parliament, and when its staff members heard about particularly problematic life-form patents at the EPO, they participated in—and sometimes led— oppositions. But by the late 1990s, it had positioned itself as an institutional watchdog and developed a routinized system to identify candidates for opposition.[94] Every two weeks, someone from KPAL visited the EPO's offices to search electronically through all of the newly granted patents and newly published applications. She used search words designed to uncover life-form patents, including "plant," "animal," and "embryo." She also did a preliminary assessment of the patents' importance, breadth, and shock value, and then highlighted a subset for further consideration. Based on their experience in challenging patents, KPAL staff chose final targets that they determined might set a legal precedent and that might provoke citizen mobilization and ensure allies among other civil society groups. They then designed an accompanying political strategy and, although they were not lawyers, they played an active role in developing the legal strategy.

One civil society activist who had been involved in many KPAL-led oppositions reflected on how the process had changed between the Oncomouse case, which was initiated in the early 1990s, and those launched in the 2000s:

> I think [in the early cases] the patent lawyers [we worked with] tried to make the full range of legal arguments, and we were not able to discuss it. So if the patent lawyer said, go for inventive step, we would go for inventive step

[questioning whether the invention established sufficient distance beyond the state of the art] because nobody really knew what inventive step was about. And then we watched it happen. So in the Oncomouse case it was more broad in the beginning; it was from other scientists, which was more or less the same thing, so there were issues of novelty, I think it's not the patent law that changed, it's that the people who run oppositions know a little more. Yes, because of getting into the procedures and listening to the cases.[95]

Initially, KPAL and its allies challenged life-form patents on any grounds, simply because they found them problematic. But as they gained expertise in the patent system, they began to use their evolving knowledge to make strategic choices that would bring them closer to their overall goal: eliminating, or at least dramatically reducing, life-form patents. By contrast, while US challengers used similar strategies—including attempts to transform political opportunities—they had not yet found a way to influence patent decision making. The barriers erected around the US system seemed impenetrable. But hESC patent challengers in the United States had achieved some attention, and as we will see in the next chapter, US critics would adapt further to the barriers and experience some success.

KPAL found the Edinburgh patent through one of its routine checks of the EPO database (the EPO would later refer to the patent as a "mistake").[96] For KPAL it seemed to be the perfect case. It would force the EPO to further define its approach to ethical issues. It would also test the BPD, which had specified the *ordre public* clause with prohibitions on "processes for modifying the germ line genetic identity of human beings" and the "uses of human embryos for industrial or commercial purposes." Both specifications seemed relevant to the Edinburgh patent. And while KPAL was not against hESC research, its staff thought that the patent might inspire considerable public interest, given the ongoing worries about the practice. So the organization initiated an official opposition. It first alerted the media,[97] which seemed quite interested to learn that the patent covered transgenic humans, and then organized a large protest in front of the EPO's Munich office (figure 4.1). KPAL and its allies frequently organized such theatrical actions to coincide with their official oppositions. These efforts demonstrated its historical connections to Greenpeace, which often used such tactics to increase the likelihood of media coverage and public attention.[98] These strategies were likely to be particularly powerful in a European context that was experiencing heated debate not only regarding hESC research, but also about the implications and appropriate regulation of

Figure 4.1. Protesters, led by Kein Patent Auf Leben, barricaded the doors of the EPO in response to the University of Edinburgh patent covering hESCs and transgenic humans. Photograph by Thomas Einberger, for Greenpeace (February 22, 2000).

biotechnology overall.[99] KPAL staff hoped that this approach would make EPO personnel consider their responsibilities to a public that they didn't usually consider.[100]

In a stroke of bad luck for the EPO, the Edinburgh patent demonstration coincided with a meeting of the EPO's Administrative Council. As a result, there was even more attention to the controversy, which generated press coverage across Europe, Canada, and Australia.[101] The EPO and its Administrative Council could have ignored or dismissed these protestors, as their US counterparts would probably have done. They could have suggested that these citizens were targeting a domain that was disconnected from ethical issues or emphasized the potential benefits of hESC research and treatment. But instead, EPO officials felt tremendous pressure.[102] One recalled,

> That was really the worst situation that we had ever experienced. Namely, we had [KPAL] coming up to us on a Monday morning at seven o'clock, breaking up the EPO so that no one could get in and out. And it was at the time a meeting of the Administrative Council, which means, the supervisory body of the EPO and as it were, at the same time, we had the EU com-

missioner visiting the EU commission office, which is in the same building. It was Monsieur Busquin, who was then commissioner in charge of research, so he was intimately associated with the whole situation, and he couldn't believe what he saw and then automatically from that moment it had a European dimension.[103]

Over the next few months, KPAL continued its public-relations campaign and also began the official opposition. With the benefit of the networks it had developed in previous cases, it assembled numerous official opponents of the Edinburgh patent, including environmental organizations, a bioethicist, religious groups, and pro-life advocates primarily from Germany and Austria.[104] It also worked with government ministries from Germany, Italy, and the Netherlands, who filed separate oppositions. In total, fourteen organizations—some of which opposed hESC research generally—challenged the patent. Of course, this contrasted with the political environment emerging in the United States, where the pro-life groups most concerned about hESC research were silent about patents. In Europe, the patent system was slowly becoming a central front in the battle over hESC research and treatment.

To provide legal representation and assistance in the Edinburgh case, KPAL enlisted British barrister Daniel Alexander. It was not easy for civil society groups to find the patent-law expertise necessary for these oppositions. A KPAL staff member observed that many patent lawyers refused to help because they would be seen as going "against the interests of patent lawyers," against their "tribe";[105] indeed, Rifkin and his allies had experienced similar frustrations in the early US life-form patent battles.[106] But even those patent lawyers who wanted to help the life-form patent challengers could cause problems, because they focused on winning the case by any means necessary, rather than the bigger strategic objectives that activists had. But life-form patent challengers had gradually assembled a small cadre of sympathetic patent lawyers to facilitate their efforts. Alexander, according to a civil society activist, was interested in the issues "from a professional perspective. He said yes, we really need to discuss these things. So he's really well-acknowledged, he has a certain perspective, which is well known, which is that patents shouldn't be broad, but as far as I can see he is still part of the system."[107] An independent lawyer for hire certified to practice in the UK, at the EPO, and in front of the EU Court of Justice, Alexander had interests and expertise related to civil society oppositions and *ordre public*.[108] He assisted Greenpeace, KPAL, and their allies in many cases, providing them with legal expertise that allowed them to make arguments that might be treated as credible and legitimate within the European patent system. But he did not seek to be the primary strategist, and these cases were only a small part of his portfolio; he provided counsel on a variety of intellectual property cases to both small and large companies.

Challenging the Morality of hESC Patents

The University of Edinburgh responded to the oppositions by quickly limiting its patent to exclude the production of a transgenic human. But opponents were not satisfied.[109] The patent still covered a process for producing human embryonic stem cells. This, opponents suggested, would both encourage a market in human embryos and encourage their destruction—which would violate human dignity. Therefore, they argued, the patent violated both the EPC's *ordre public* clause, and the BPD's explicit language that forbade patents on inventions that involved "the uses of human embryos for industrial or commercial purposes."[110] This clearly linked life-form patent challengers and the hESC research critics. Both worried about the impact of biotechnology on human dignity and agreed that the commodification of hESCs through patents would devalue human life. Because the patent system had already defined patents as moral objects, they suggested, it had to accept a role in governing hESC science and technology.

Two aspects of this framing are worth noting. First, European challengers rallied around the right to human dignity, which had arisen in the BPD and had a long history in European law, as discussed in chapter 2. This was different from US concerns regarding hESC research, which focused on the sanctity of human life. But whereas "sanctity" focused implicitly on matters of life and death, the European focus on "dignity" covered the process of creating hESCs (which involved embryo destruction) as well as commodification and exchange in the marketplace. Second, in Europe, even patents on the *process* of creating hESCs were treated as potentially violating human dignity. In the United States, by contrast, even patents on the hESCs themselves had not provoked questions about the relevance of the 1987 Quigg memo, which prohibited patents on "human beings."

Unlike the Oncomouse opposition, which relied on scientific articles, public opinion, and national laws, the Edinburgh patent opposition drew heavily on what we might classify as ethical expertise and reasoning. In even greater volume than in the Oncomouse application, opponents submitted scholarly and media articles related to the ethics of research and patents on stem cells and cloning, along with statements from experts in moral and political philosophy. This would test how the EPO evaluated morality. In the Oncomouse case, it had defined moral issues narrowly and relied on legal interpretation, but now opponents were suggesting that ethics and ethicists had a more formal role to play. KPAL, for example, submitted a report from the Danish Council on Ethics entitled "Patenting Human Genes and Stem Cells," which stated, "The members of The Danish Council of Ethics consider all forms of stem cell patenting problematic. The members based this view first and foremost on the argument that patenting stem cells involves commercializing and commodifying the human organism, which cannot be combined with respect for the dignity of mankind."[111] To the Danish Council, which included lawyers, medical professionals, religious figures, social workers, and a journalist, the patent system was a relevant and important regulatory space that shaped markets and the definition of commodities. Furthermore, the role of the government was to set market rules with the moral community in mind. Neither the market nor the patent system was exempt from prior ethical consideration. While the Danish Council recognized significant benefits in conducting hESC research, and ultimately did not suggest outlawing stem cell patents entirely, it recommended extreme caution and strong limits.[112]

Opponents submitted other kinds of evidence as well. As in previous cases, they tried to justify their position using public opinion. They submitted more than thirty-six hundred signatures from citizens in Switzerland, Austria, and Germany who wanted to join the patent challenge on moral grounds.[113] The Dutch and German opponents discussed their national laws, which outlawed the types of manipulation required to produce the invention; the legislative representatives of the Dutch and German people, they argued, had already found the embryo manipulation involved in producing hESCs to be morally problematic.[114] Although the EPO had already ruled in earlier oppositions that it saw public opinion and national laws as poor indicators of *European* public order and morality, these submissions allowed opponents to show the EPO that citizens across Europe were scrutinizing—with significant moral concerns—the bureaucracy's activities.

The European Parliament also challenged the EPO's moral calculation regarding the Edinburgh patent: 285 of its members condemned the decision in a nonbinding resolution in December 2000.[115] (Of 626 members at the time, 131 voted against the resolution, and 7 abstained.) The resolution argued that "no consideration of research, and still less of profit, can be allowed to override that of the *dignity of human life* [emphasis added]," and it called "for this principle to be written into the Treaty on the European Union in the future."[116] In support, the resolution cited the UN's Universal Declaration on the Human Genome and Human Rights, which articulated a "right" to respect human dignity.[117] This language was notable. The European Parliament had accepted opponents' characterization that patents shaped human dignity, and suggested that upholding this value was more important than maintaining the interests and freedoms of patent holders. In addition, it characterized the patent system as part of a broader policy apparatus contending with the ethical implications of a particularly controversial area of science and technology. Finally, the European Parliament resolution implicitly suggested that such ethical concerns could not be left to scientists and entrepreneurs. These researchers were potentially self-interested; therefore, government, including the patent system, had to take on moral responsibility. Whereas in the United States the debate over hESC patents suggested that hESC innovation was beneficial and that the role of decision makers was to facilitate it, in Europe decision makers expressed more ambivalence toward both innovators and hESC innovation itself.

The resolution also rebuked the EPO's decision-making approach. It demanded "a review of the operations of the EPO to ensure that it becomes publicly accountable in the exercise of its functions, and to amend its operating rules [so that it can revoke] a patent on its own initiative."[118] To the Parliament, an "accountable" EPO would explicitly consider values, including human dignity, and accept regulatory responsibility over innovation. It would also recognize the moral complexity of innovation and accept responsibility in adjudicating this complexity.

By intervening in this way, approximately half of the MEPs sent the message that they took active responsibility for patent-system decisions and expected the EPO to do so as well. They also expected the patent bureaucracy to interpret the BPD, which parliamentarians had passed two years before, broadly in order to consider moral issues. Serious consideration of patents' moral status, the European Parliament seemed to conclude, required a bureaucracy that had greater authority and was accountable to the public in ways beyond achieving procedural objectivity. Again, this was a very different vision of innovation, patents, morality, and bureaucratic power than the one that had emerged in the United States. Research on hESCs was seen by the vast majority of patent-system decision makers in the United States to be morally good. On the rare occasion that a congressman tried to bring in moral concerns specifically related to the patentability of human beings, the subject had been quickly abandoned. And while there was growing evidence that patents might hurt research, these concerns were largely dismissed by referring to the moral authority of patent holders, who were seen as achieving the public interest by operating in their own interest.

In its official reply, the University of Edinburgh argued as the EPO had in the early discussions over life-form patents: that *ordre public* was extremely limited and therefore not relevant to the case: "The Exclusions to patentability in Article 53(a) EPC [the *ordre public* clause] are intended to prevent grant of patents that would induce rioting, terrorism, or that would provide serious public offence. This is not the case in the present patent. The exclusions are also to prevent grant of patents that are contrary to public policy (*ordre public*) but if the subject matter of the patent is encouraged and supported by public policy in some member states then patentability cannot be denied. Public policy in some member states does support therapies based on embryonic stem cell research."[119] But the European patent system and the EPO itself had already shifted their understanding of both *ordre public* and the moral status of patents. So when the EPO decided to hold oral proceedings and consider concerns related to *ordre public* in late 2001, the university had to adjust. It changed tactics and simply argued that while it recognized the relevance of *ordre public*, its patent did not violate the clause.[120] It had, in other words, accepted that the patent system was an appropriate place to deal with moral concerns regarding hESC science and technology.

To make its arguments, the University of Edinburgh marshaled both legal and ethical evidence. It disputed the idea that the European Patent Convention or the BPD forbade patents on the process of making human embryonic stem cells. The BPD, it argued, said nothing about hESCs. It also invoked the still ongoing Oncomouse case, noting that the Opposition Division had concluded that controversial inventions could be patented if they had substantial medical benefit: "In the face of indisputable medical benefits associated with the invention, hypothetical potential risks were not a ground for denying patentability on ethical grounds. In particular, it was held inappropriate to deny patentability on the basis of possible risks in the absence of conclusive evidence. Moreover, it was not the place of the EPO to take on the roles of various regulatory authorities."[121] But referring to the Oncomouse case actually demonstrated the shifting grounds of the European patent system. By the early 2000s, the EPO and the European patent system more generally were starting to envision themselves as performing important regulatory functions.

The University of Edinburgh also cited the conclusions of the European Group on Ethics, which had stated that "it is ethically acceptable to patent human stem cell lines that have been modified . . . and to patent methods involving human stem cells from any source." It reported further that "the EGE observed that moral opinions on patenting human stem cells are diverse and that the 1998 Biotechnology BPD [*sic*] does not specifically exclude the patenting of cells from human embryos."[122] The university hoped to convince the

EPO that these ethical issues were irrelevant, but if they were considered, then the potential medical benefits should outweigh the problems with commodifying and destroying human embryos.

At the oral proceedings convened to discuss the Edinburgh patent opposition in July 2002, KPAL asked Ingrid Schneider, a political scientist from the University of Hamburg whose research focuses on bioethical issues, democracy, and the public interest, to speak on its behalf.[123] Greenpeace Germany invited Dietmar Mieth, a professor of theological and social ethics from the University of Tuebingen who sat on several German ethics committees, to testify. (In order to qualify as "experts" who can testify in the oral proceedings, the opponents must get the EPO's approval in advance by notifying them and providing the proposed expert's CV. However, there is no evidence that EPO vets these requests in any way; the patentee and opponents simply have to register those they plan to use as experts.) Unlike the US bioethicists involved in patent discussions, who focused on the benefits of hESC research, the professors testifying at the EPO focused on issues of commodification. As they did so, they continued to carve out a role for this type of ethical expertise in the patent bureaucracy and to emphasize the European patent system's moral responsibility in terms of human and animal dignity.

In July 2002, the EPO's Opposition Division decided that hESCs, and the processes of making them, were unpatentable due to *ordre public*.[124] As it did this, it further articulated its responsibility in shaping both European innovation and the market, and defined its understanding of ethics further. It based this decision on legal precedent and interpretation, as it had done in the Oncomouse case. It argued that hESCs were not patentable according to Article 6 of the European Union's BPD, which interpreted the *ordre public* clause as disallowing patents on inventions that used human embryos for industrial or commercial purposes, particularly when read in the context of the "recitals" found in the introduction to the BPD (which the EPO argued demonstrated legislative intent). Recital 16 states that "the human body, at any stage in its formation or development, including germ cells, . . . cannot be patented"; Recital 38 states that "processes, the use of which offend against human dignity, such as processes to produce chimeras from germ cells or totipotent cells of humans and animals, are obviously also excluded from patentability," and Recital 42 provides a very narrow exception for "inventions for therapeutic and diagnostic purposes which are applied to the human embryo and are useful to it."[125] These recitals did not have the force of law, and could have been essentially dismissed. But, perhaps because it was well aware of the enormous public interest in the case,[126] the Opposition Division decided to use a broad interpretation of the BPD rather than the narrow one favored by patent applicants, which considered only the text of the legally binding Articles.

The Opposition Division considered, but rejected, the other legal evidence that the opponents and patentee had introduced. It noted that the EPO had already ruled in a previous decision that *ordre public* should be defined in European, rather than national, terms: "The concept of morality is . . . related to the belief that some behaviour is right and acceptable whereas other behaviour is wrong, this belief being founded on the totality of the accepted norms which are deeply rooted in a particular culture. For the purposes of the [European Patent Convention], the culture in question is the culture inherent in European society and civilisation."[127] Therefore, national laws either promoting or opposing stem cell research were not relevant. Similarly, they dismissed the thousands of signatures submitted by the opponents because they were not an appropriate reflection of Europe as a whole.

It did not, however, automatically dismiss the ethical knowledge and expertise invoked by the opponents or by the patentee. For the first time, the Opposition Division acknowledged that the BPD had established the authority of the European Group on Ethics in Science and New Technologies as an expert body relevant to the ethics of patents. This was significant, because it demonstrated an acceptance of the relevance of ethics expertise to EPO decision making and the legitimacy of this ethics advisory body in particular. But, the Opposition Division argued, the EGE's report on the matter was illogical and inconsistent with the general principles of patent law.[128] The EPO could have used the EGE's recommendations to justify a decision in favor of hESC patents, but it chose not to do so, demonstrating an evolving understanding of its moral responsibility and its duty to be publicly responsive. Given the EPO's history and expertise, its reliance on legal interpretation is not surprising. But in many respects, its decision was. Rather than taking a position aligned with its traditional organized interests, it chose to interpret the law in a way that might strengthen its legitimacy among citizens, national governments, and the European Parliament. In the process, it validated the relevance of explicit ethical expertise and discussion.

Over the next few years, the Edinburgh case made its way through the appeals process. In the meantime, the Wisconsin Alumni Research Foundation filed its EPO application for the patent covering primate embryonic stem cells and methods of making them, that had stimulated controversy over the impact of patents on research in the United States. The EPO's examiners immediately rejected this patent on the basis of the Opposition Division's Edinburgh patent decision.[129] The WARF appealed the decision, sending it to the EPO's Technical Board of Appeal. In November 2005, the TBA referred the case to the highest judiciary body in the organization, the Enlarged Board of Appeal.[130] The EBA was made up of both experienced patent examiners and patent lawyers. By this point, the patent had mobilized an even larger group of critics than the Edinburgh patent had, including religious organizations, social scientists, bioethicists, patent lawyers, and government ministries. Together, they submitted thousands of pages of amicus briefs (with hundreds of thousands of signatures) against the WARF patent. Most of these briefs, like the oppositions in the Edinburgh case, made ethical or rights-based arguments. All argued that the BPD clearly prohibited the hESC patents.[131]

The briefs focused on the moral problem with creating a market for hESCs. An Italian professor of economics and statistics suggested, for example, that the moral harm of commodifying life forms outweighed the economic interests of the applicants.[132] These interests, he argued, prevented patent holders from having moral authority. Others offered a rights-based critique. The European Center for Law and Justice, a pro-life human rights advocacy group, noted that the WARF patents should be prohibited because "the wealth of protections provided to human dignity and the protection of European subsidiarity by European Union law and policy statements, while not binding on this esteemed Board, nonetheless provides a persuasive hermeneutic from which to analyze the WARF application."[133] The ECLJ also argued that turning hESCs into patentable inventions would make them commodities, which would violate human dignity and therefore, the EU's Charter of Fundamental Rights and the European Convention of Human Rights.

The president of the EPO, Alain Pompidou, weighed in as well.[134] While it was rare for the president to get involved in individual cases, this debate had generated enormous attention; the biotechnology industry, a variety of advocacy groups, and even average citizens were now watching the proceedings closely. Pompidou interpreted patent law as prohibiting hESC patents, and he went even further to emphasize the moral dimensions of innovation and the EPO's moral responsibility, stating that "the granting of a patent is often perceived to be an official endorsement of or reward for a particular invention." As with the European Parliament's hESC patent resolution, this demonstrated a completely different view of innovation, patents, and their governance than we saw in the United States. After all, in *Chakrabarty*, the US Supreme Court had argued the opposite, that patents only determined "whether research efforts are accelerated by the hope of reward or slowed by want of incentives, but that is all."[135] And in the US hESC patent debates, decision makers and organized interests had assumed the moral benefits of hESC research. But Pompidou saw innovation as having both positive and negative implications, and patents as essentially providing a moral stamp of approval. As a result, the EPO had an active role to play.

Supporters of the WARF patent, including patent lawyers and representatives of the UK and European biotechnology industries, also submitted amicus briefs.[136] As in the Edinburgh case, they invoked ethical, scientific, and legal knowledge that supported their position. The UK Biotechnology Industry Association argued that patents were "fundamental in fostering motives necessary for the advancement of science that will bring medical and therapeutic benefit and the eventual improvement of the quality of human life," citing reports from the Centre for the study of Bioscience, Biomedicine, Biotechnology, and Society (BIOS) at the London School of Economics and the University of Nottingham's European Patent Law and Ethics group.[137] As with the patent applicants in the Edinburgh case, rather than rejecting moral discussion completely, the WARF brought its own army of experts to bear on the issues. In this respect, the debate over ethical knowledge and expertise in the European patent system had begun to look no different than the science policy controversies that so many scholars of science and technology studies have analyzed, with each side bringing in evidence to support its position.[138] The WARF side used its evidence to emphasize the moral authority of innovators and the benefits of innovation.

In 2009, the Enlarged Board of Appeal upheld the Examining Division's initial decision and prohibited the WARF patent. Through its decisions in both the Edinburgh and WARF cases, the EPO had decided that hESCs, and the processes of making them, were unpatentable. In making this decision, it reiterated a connection between patents and morality, emphasized its own ethical responsibility to take proactive measures to limit the commodification of life forms, and defined its role as part of the apparatus governing hESC research.

EPO officials saw the Edinburgh case as another important turning point in their understandings of both the institution's role and its sources of political legitimacy. One official reflected on the history of the case:

> There came this strong development that civil society all of a sudden took part in the process, which was usually thought to be happening in an ivory tower. That ivory tower had a crack and we had to react, especially after there was condemnation by the Parliament, the European Parliament. That our president was questioned by the national governments and also the

member states were growing a bit quiet. . . . I mean, we do make mistakes. We do grant patents erroneously, *but there's a difference whether you make an error in toothpaste technology or genetic engineering* [emphasis added].[139]

Citizens with concerns quite different from the EPO's traditional stakeholders had begun to establish themselves as important constituents of the European patent system, and both the European Parliament and the EPO had taken notice. It was no longer adequate for the EPO to maintain its legitimacy on the grounds that it was an objective, technical bureaucracy separate from the political arena. In other words, while officials at the PTO would likely argue—and often did, using slightly different examples—that patent applications on toothpaste technology and genetic engineering should be treated in an identical manner, to EPO officials they were different because of the moral issues involved.

In the following years, the EPO tried to implement this new understanding of its role and responsibility in a number of ways. It began to publish policy reports on issues that lie at the intersection between patents, technology, and society and to participate in technology policy discussions inside Europe and beyond.[140] It issued press releases and fact sheets when it granted a patent that was likely to become controversial, regarding opposition and appeals proceedings on controversial patents, and in response to general areas of controversy (e.g., patents on living organisms or software).[141] These documents described the issues of public concern and the reasons behind the bureaucracy's decision making. These public-relations efforts were overseen not only by the Communications Department, but also by Issue Management Groups (IMGs), composed of examiners, patent lawyers, and public-relations specialists, all from inside the organization.[142] Each IMG specialized in one controversial technical area (e.g., computer-implemented inventions, biotechnology) and helped communications officials develop strategies to deal with patent applications that had been identified through the Sensitive Cases system, while also helping them interact generally with stakeholders on issues that had become controversial. IMG members were often sent out as envoys to patent-related events throughout Europe, so that they could listen to the concerns of various groups and also explain the position of the office. They also might review drafts of press releases or speak to members of the EPO's Administrative Council or the European Union's Parliament to alert them to the impending issuance of a controversial patent. The IMGs and increased contacts with the press were developed to better communicate with the public and explain the EPO's decisions, and also to gather information about issues of public concern for further discussion inside the EPO.

Validation in the Courts

As the Edinburgh and WARF patent oppositions crawled along, Christoph Then (this time on behalf of Greenpeace) also challenged hESC patents through the national courts. Eventually these challenges arrived at the EU's highest court, the European Court of Justice, which was potentially another source of pressure on the European patent system. In 1999, two years before the EPO granted the Edinburgh patent, the German Patent Office awarded Oliver Brüstle, a researcher at the University of Bonn and a vocal proponent of hESC research, a patent covering methods of deriving neural cells from hESCs.[143] Greenpeace immediately opposed the decision, arguing that the patent violated the BPD. The German Patent Office revoked the patent; Brüstle filed for and received a modified patent; and Then and Greenpeace sued in 2004.[144] The German federal court ruled in favor of Greenpeace, and Brüstle appealed the case to the German Supreme Court. The German Supreme Court referred the case to the ECJ, seeking clarification on several aspects of the BPD, including the meaning of "human embryos," the meaning of the expression "uses of human embryos for industrial or commercial purposes" in the BPD (which by then was also figuring prominently in EPO's decisions to deny the WARF patent and limit the Edinburgh patent), and whether the destruction of the human embryo matters for determining patentability if it is not explicitly discussed in the patent's claims. Although Brüstle argued that the case jeopardized the future of hESC research in Europe, Then emphasized that his focus was on the commercialization of human embryos: "Even if research on embryos is to be allowed, it's not right to commercialize the process. . . . We believe there should be a clear separation between research and patenting products."[145] To Then, the patent system had an important and distinct role to play in the governance of hESC research. Focusing regulatory attention on the research enterprise would not address the patents' socioeconomic impacts or the moral impacts of turning hESCs into commodities.

In October 2011, the ECJ agreed with Greenpeace and echoed the EPO's position. In its first decision on the patentability of life forms,[146] Europe's highest court defined a human embryo as any fertilized human ovum in which the "process of development of a human being" was taking place, and read the BPD to exclude patents on products and processes related to hESCs because they involved the "commercialization or exploitation" of a human embryo. It also decided that hESCs were unpatentable if they involved the destruction of an embryo, even if that destruction was not explicitly discussed in the patent claims. In making this decision, the ECJ relied heavily on the text of the BPD, especially the language in Article 6 that provided the basis for the EPO's decisions in the Edinburgh and WARF cases.

The ECJ also reinforced the connection between the European right to human dignity and patent law, and therefore the patent system's moral responsibilities. As discussed previously, while challengers had made this connection in the debates over the Edinburgh and WARF patents, the EPO had not justified its decisions explicitly in moral terms. But, the ECJ did, citing the Preamble to the Biotech Patent Directive as proof that "the European Union legislature intended to exclude any possibility of patentability where respect for human dignity could thereby be affected." It noted further that the "use of biological material originating from humans must be consistent with regard for fundamental rights and, in particular, the dignity of the person. Recital 16 in the preamble of the Directive, in particular, emphasizes that 'patent law must be applied so as to respect the fundamental principles safeguarding the dignity and integrity of the person.'"[147] To the ECJ, hESC innovation could jeopardize human dignity, patents were moral objects, and the patent system's responsibility was to privilege human dignity over the right to research and the potential social benefits of hESC research.[148]

Not surprisingly, many stem cell scientists across Europe were deeply upset by the ECJ's decision, arguing that it would create a hostile environment for hESC research.[149] In its decision, the ECJ had embraced an active moral responsibility for the European patent system, and determined that upholding the right to human dignity warranted active engagement, even in the form of limiting patentability.

CONCLUSION

The US and European hESC patent controversies reinforced rather different pictures of the patent system's role and responsibility in governing controversial science and technology. Along with these different views emerged very different political environments, preoccupied with different policy questions and debates, and populated by different organized interests who had different concerns. In the United States, where patents had been established as techno-legal objects to be governed by a narrow set of experts and by institutions focused on procedural and legal objectivity, the patent system was now far removed from concerns about the morality of hESC research and treatment. Even critics of hESC research generally seemed to assume that the patent system was not the place to govern the field's moral implications, including the commodification of embryos or hESCs. But patents were still controversial, perhaps because scientists were already concerned that the ethical controversy would limit their freedom to do research. These worries elicited more attention from decision makers than previous bioethics concerns, likely because they suggested implicitly that patents on research tools could stifle innovation and interfere with the development of new products for the marketplace. Decision makers were still reluctant, however, to entertain the idea that patents stifle research and to take steps to address potential problems. Influenced by arguments that patents were needed for innovation and industry, they sought clear proof of negative impacts before they changed policy.

Europe, by contrast, had dealt with the potential impact of patents on research long ago, taking a proactive approach. They had made a very different calculation than their US counterparts had made, who deemed even a small risk of stifling research as too great. With regard to hESC patents, the European debate focused on bioethical implications—specifically, on the implications for human dignity. After all, civil society groups had achieved some success in the Oncomouse case, and the EPO's bureaucratic culture seemed to be changing. As the oppositions proceeded and gathered support, it became clear that national governments and European citizens now expected the patent system to participate in the governance of ethically controversial science and technology. It was responsible for protecting life forms and making moral judgments about what should be permitted in the marketplace. Over time, European patent-system institutions, including the European Court of Justice and the EPO, accepted this responsibility and changed their activities accordingly. These changes were based on the idea that patents had significant ethical implications and served as official endorsements of particular areas of research.

The controversies also revealed different understandings of moral authority in the US and European innovation systems. Despite the vigorous debate over the ethics of hESC research that was taking place in the United States, its patent system characterized hESC innovation as fundamentally ethical due to the health and social benefits it might produce. While patents themselves continued to be understood as techno-legal and therefore amoral, the debate reinforced the idea that both patent holders and researchers were fundamentally ethical and should be trusted to operate sensitively and with the public interest in mind. Despite growing concerns about how patents could stifle research, there was skepticism of the idea that patent holders might operate against the public interest. And while the conflict over whether patents might stifle innovation created a potential conflict between a "good" scientist and a "bad" patent holder, decision makers and traditional patent-system interests resolved it—at least temporarily—by suggesting that it was simply false. Patent holders, like scientists, behaved ethically because they increased innovation and expanded markets. Those who did not were characterized as "bad apples." They were so rare that they did not warrant additional regulations in the United States: most patent holders would license their inventions widely, which would stimulate innovation and benefit society, because it was in their own best interests. This characterization implied considerable trust in both scientists and patent holders, and distinguished the patent system as an inappropriate space for governing issues related to the morality of research and technological development. In the European patent system, by contrast, innovation was seen as morally ambiguous, and innovators were morally complex. While they produced contributions that had social benefits, their work also produced ethical and social problems that they might be too self-interested to see. This created an important responsibility for European patent-system institutions, which were viewed as part of the broad apparatus for governing ethically controversial science and technology.

5 | HUMAN GENES, PLANTS, AND THE DISTRIBUTIVE IMPLICATIONS OF PATENTS

While the hESC patent debates were unfolding, new frustrations regarding life-form patents emerged as biotechnology began to deliver medical and agricultural products to the marketplace. Patent holders were wielding their ownership rights aggressively in both the United States and Europe, generating negative socioeconomic implications not just for scientists but for average citizens. Public-health professionals, physicians, and patients observed that owners of human gene patents were establishing monopolies on genetic testing, which allowed them to charge high prices for access. These patents also lowered incentives, they argued, to improve the technology. Farmers complained that plant patents were transforming traditional agricultural practices in deleterious ways, making farming very expensive and limiting food choices for consumers. This emphasis on the distributive implications of life-form patents evolved just as "access to knowledge" and "right to health" movements were gaining strength in the developing world. In those countries, civil society groups argued that patent-based monopolies were exacerbating economic inequalities by limiting access to technology.[1] They demanded that governments rethink their participation in international patent and trade agreements and also create exceptions, including compulsory licensing, in their patent policies that would provide flexibility to consider public-interest concerns.[2] Some governments seemed to be listening.[3]

How did the United States and Europe respond to these kinds of concerns? If we focus solely on the legal outcomes, we might conclude that neither the US nor the European patent system seemed particularly responsive. The European patent system, including the EPO, which we might have assumed would be more sympathetic, allowed patents on both human genes and plants in the face of civil society opposition. The United States rejected challenges to plant patents, and although it ultimately prohibited patents on human genes, it did so because the Supreme Court deemed them "products of nature." Considering these outcomes, we might decide that the comparative differences I have unraveled throughout this book are not so important after all.

155

But if we step back and look at the human gene and plant-patent controversies in broader political and institutional context, a much more coherent comparative picture continues to emerge. European patent-system institutions did in fact recognize responsibility for the distributive impacts of patents and took both formal and informal steps to address these concerns. Meanwhile, their US counterpart had developed its laws, institutions, and political environment in such a way that it was virtually impossible for distributive concerns to be taken seriously. The long-standing differences in how the two jurisdictions saw patents—as techno-legal in the United States, and as moral and policy objects in Europe—coupled with decades of debate over life-form patents, actually had significant long-term implications for the two patent systems.

UNITED STATES

Human Gene Patents and the Right to Health

Human gene patents first generated criticism in the United States in 1995. In keeping with his previous efforts, Jeremy Rifkin assembled two hundred religious leaders from various faiths to raise concerns regarding the commodification of human life. A United Methodist bishop noted, "We're saying let's treat genes the same way as chemical elements. . . . Patenting of life reduces it to its commercial value. When ownership of life becomes a commercial commodity, its worth depends upon its marketability. . . . That which is profitable may not be that which is beneficial."[4] This campaign fizzled quickly, capturing even less public and policy attention than previous moral challenges to life-form patentability.

In 1996, Rifkin tried again. This time, he focused on the PTO's imminent decision to issue patents on the BRCA1 gene to Myriad Genetics, a biotechnology company based in Salt Lake City, Utah (later, the PTO would also issue patents on the BRCA2 gene to Myriad Genetics as well as to UK researchers).[5] Although the BRCA genes only conferred increased disease susceptibility— they are linked to 5–10 percent of breast cancers and 15 percent of ovarian cancers, and confer a lifetime risk of up to 65 percent for breast cancer and up to 40 percent for ovarian cancer—researchers and physicians saw the gene discoveries as important steps forward in understanding and preventing breast and ovarian cancer.[6] Rifkin and his allies, including womens' groups, cancer patient advocates, and biotechnology watchdog organizations, questioned the distributive impacts of patents on these BRCA genes. The gene discoveries had been heralded across the world as major steps forward for patients and those concerned about their cancer risks.[7] But patent critics argued that the patents would restrict research on the genes as well as access to related genetic testing and therapies. Wendy McGoodwin, the executive director of the Council for Responsible Genetics, a public-interest organization focused on issues related to genetics and society, observed, "It seems unfair that a private company using the benefit of [an international scientific collaboration] would patent it, then turn around and charge us more."[8] But this effort disappeared quickly as well, despite the similarities between these concerns and the protests over cDNA patents coming from the scientific community at that time.

Soon, Myriad Genetics and other owners of human gene patents would validate the Rifkin coalition's concerns.[9] In 1997, Myriad used its patent position to shut down almost all other providers of BRCA testing. This included anyone who was returning test results to patients, even in the context of research.[10] The company claimed that its gene patents prohibited anyone without a license from offering BRCA gene testing, regardless of the method used or its connection to research—and it chose not to sell broad licenses for its patents. One competitor did challenge Myriad's patent claims, but quickly gave up when it became clear that Myriad would defend its patents strongly and was not open to negotiation.[11] Myriad thus became a virtual monopoly in the United States, and it charged a relatively high fee for its services: its comprehensive test cost approximately three thousand dollars. At the time, clients often paid out-of-pocket to maintain their privacy, concerned about discrimination from their employers and insurance companies.[12] Impressed by Myriad's success, other gene patent holders adopted similar patent strategies. This made diagnostic laboratories and scientists reluctant to do research on particular genes and provide testing because they were worried about patent infringement.[13]

Health care professionals, scientists, and patient advocacy groups were frustrated that these patent holders were making genetic testing so expensive that it limited access,[14] but most kept quiet initially. Some concluded that it was socially and politically unacceptable to articulate their worries, concerned that they would have little support among their peers or that testing companies would seek retribution.[15] After all, the Rifkin coalition had achieved little attention. On the rare occasion when they did speak, these patients and scientific and medical professionals focused on individual patent holders, rather than patents themselves, as the problem. They seemed to accept the prevailing idea that patents were not the problem, but that patent holders occasionally might be. And they also appeared to view the patent system as generally beneficial and necessary, because it stimulated innovation.[16]

However, as I discussed in the last chapter, scholars and policymakers were slowly taking notice of these emerging frustrations.[17] And, as concerns about whether patents might hurt research began to appear on the policy agenda, so too did worries that they might affect "access" to and the "quality" of health care. The Secretary's Advisory Committee on Genetic Testing (SACGT), convened by the Department of Health and Human Services (HHS), invited some of the lawyers and social scientists examining gene patenting to testify on the topic in June 2000. The SACGT was part of a long legacy of national expert advisory committees focused on the ethical, health, and social impacts of genetics and biotechnology.[18] But after this hearing, it simply sent HHS Secretary Donna Shalala a letter suggesting that she "consider whether further study of the issues is warranted."[19]

The issue was far from dead. Around this time, the NIH asked the National Research Council (NRC) to study the impacts of patents on biomedical research and public health, and many of the same researchers in this academic subfield testified. Then, in 2002, the new HHS secretary asked the SACGT's successor, the HHS Secretary's Advisory Committee on Genetics, Health, and Society (SACGHS), to consider the "impact of patent policy and licensing practices on access to genetic technologies" in its founding charter.[20] Although it is not entirely clear why the secretary asked the SACGHS to examine the impacts of gene patents, especially given the long legacy of dismissing a link between patents and the implications of patent-based monopolies, it seemed that the growing body of scholarly research as well as anger with Myriad Genetics had had an impact. Myriad's monopoly focused on a common and high-profile disease; it affected a large number of health care professionals, scientists, and patient advocacy groups; and some were complaining publicly.[21]

The SACGHS, which included medical and molecular geneticists, entrepreneurs, lawyers, and public-health professionals, took up the issue in 2006 and conducted the most comprehensive study to date of the impacts of gene patenting on research and health care. It began its deliberations after the NRC issued its report on the matter, which had concluded that policymakers should consider patent exemptions for the purposes of research and public health.[22] The SACGHS discussed the issue during many meetings, heard testimony from several experts, solicited case studies, and reviewed public comments. Finally, in March 2010, it issued a report that drew provocative conclusions that went far beyond the findings of any previous policy committee, including the one convened by the NRC. In a letter to HHS Secretary Kathleen Sebelius, the committee chair stated that "patents on genetic discoveries do not appear to be necessary for either basic genetic research or the development of available genetic tests. The Committee also found that patents have been used to narrow or clear the market of existing tests, thereby limiting, rather than promoting availability of testing."[23] The committee issued a series of recommendations to address these problems, including "exemptions from infringement liability" for both those doing genetic research and those offering genetic testing. The SACGHS had established patents as playing a central role in shaping the socio-economic implications of genetic testing, drawing a connection that the US patent system had long resisted.

It should come as no surprise by now that both patent lawyers and much of the biotechnology industry responded negatively.[24] They could not afford to ignore the report, given the expert committee's status and the high-level policy attention it might receive. HHS Secretary Sebelius had, after all, responded by noting her "concern for the development of, and equitable access to, clinically useful genetic tests."[25] But the patent-system stakeholders responded to these experts just as they had to earlier generations of challengers, arguing again that challengers didn't understand the system and would create intolerable uncertainties if their recommendations were implemented.[26]

Whereas in the past Congress might not have stepped in at all, or might have convened hearings that reasserted the distance between patents and access to technology, the SACGHS issued its report just as the Congress was considering patent-reform legislation. And Democratic Congresswoman Debbie Wasserman Schultz from Florida, a breast cancer survivor who had tested positive for a BRCA mutation, sat on the Judiciary Committee that was reviewing it. She added language to the pending legislation based on the SACGHS's recommendations, but articulated a much narrower exception than the SACGHS had suggested.[27] She created an exception to human gene patents to allow for confirmatory diagnostic testing. This solution defined the problem as not just one of access, but one of quality: if a citizen was forced to rely on a single source for test results because of a patent-based monopoly, it could cause problems for her health care. Other than the atomic energy legislation of the 1940s, this would be the first time that Congress passed any kind of patent-law exception. But in part because of a pending court case challenging the patentability of human genes (which I discuss in further detail below), and in part because of the strong reaction from the patent-law and biotechnology communities, the final legislation (known as the America Invents Act, passed in 2011) simply asked for another review of the impacts of gene patents on testing and an assessment of how independent, confirmatory testing might be provided—to be conducted this time by the PTO.[28]

The PTO complied, but the process would be an uncomfortable one. After all, for decades the bureaucracy's personnel had dismissed the idea that the system had implications for technology policy beyond stimulating innovation and contributing to economic growth on a broad scale, and its rules and processes reflected this understanding. The bureaucracy began with two open hearings in 2012 convened by Deputy Undersecretary for Commerce for Intellectual Property Teresa Rea, which included expert testimony, discussion, and public comments. At the beginning of each hearing, Secretary Rea framed the discussion by reiterating the PTO's traditional approach to patents and governance. She argued that the patent system was beneficial to medicine overall and feared that any policy intervention would establish "a false dichotomy" between incentives to innovate and adequate access to health care.[29] To her, the most socially beneficial role for the patent system would be to continue issuing patents. This would, she suggested implicitly, increase the availability of—and therefore access to—new medical technologies. To do otherwise would be to interpret both the system and the public interest incorrectly.

The PTO heard testimony from patent lawyers, representatives from genomics companies, researchers who analyzed the link between gene patents and genetic testing, and plaintiffs from the ongoing legal case. Most of the witnesses simply dismissed the idea that patents had distributional effects, continuing to define patents as extremely limited. They characterized deleterious monopolies as occasional occurrences that were the fault of irresponsible patent holders rather than the patents themselves. Witnesses policed patent-system boundaries by making the familiar argument that those concerned about distributional effects didn't understand how the patent system worked. A representative of the American Intellectual Property Law Association observed, "Insufficient knowledge about patenting and licensing of such tests, about the relationship between genetic patents and product commercialization, and about the complexity of the genetic diagnostic business can lead to misunderstandings and misconceptions."[30] Many of them also pointed out that the SACGHS report, which had concluded that patients had limited access to genetic testing, should be taken with "a grain of salt" because it had generated "controversy" and "very public dissents."[31] Kevin Noonan, a patent lawyer who specialized in biotechnology and had submitted several briefs to the Court of Appeals for the Federal Circuit and Supreme Court on the subject, stated simply that "there's very little real evidence that patents have prevented anyone from enjoying the benefits of the new genetic technology."[32]

Most witnesses also argued that the patent system was simply the wrong place to consider the problem of access to genetic testing. Rather, they suggested, this was the responsibility of policy domains focused on health care. One witness, a patent lawyer who also teaches intellectual property law at New York University's Polytechnic Campus, made this point clearly.

I submit that many considerations of Section 27 of the [America Invents Act] and of the Federal Register notice such as 'the . . . level of medical care,' 'the interpretation of test results,' 'the performance of testing procedures,' 'the cost and insurance coverage . . . of genetic diagnostic tests,' and 'quality of care' are beyond the Constitutional remit of the Patent Office. Such matters, to me, are not proper considerations of the Patent Office, but rather are proper considerations of the Department of Health and Human Services and the Food and Drug Administration, and to me such considerations are not patent law issues. They are health care or health reform issues.[33]

As generations of patent-system decision makers and organized interests had done before him, this witness reinforced the idea that the patent system was a limited domain focused on technical and legal questions to stimulate economic growth. It was not tasked with governing health care; other domains and bureaucracies had this responsibility. The socioeconomic impacts that Congresswoman Schultz, the NRC, and the SACGHS (not to mention the earlier generation of women's and breast cancer advocacy groups) had raised were simply not under its purview. With this statement, this witness rejected the idea that patents could have socioeconomic implications. Any access concerns were related to the structure of health care provision in the United States rather than to the patent system. Finally, he framed his argument in Constitutional terms and therefore associated the current system with the very foundations of the country. To even accept the idea of patent-based monopolies, he suggested, violated the intent of the founding fathers as well as a long legal legacy.

This rejection of the distributional effects of patents and enforcement of the distance between the patent system and other domains was also clear in structural ways. Consider, for example, how the PTO organized and displayed the written comments it received for this gene-patenting analysis. A variety of individuals and organizations submitted comments, from the National Society of Genetic Counselors to the American Medical Association. But when it presented these comments on its website, the PTO put them into the traditional categories it had devised for public comments on other proposals: "Intellectual Property Organizations," "Academic and Research Institutions," "Companies," and "Individuals."[34] Comments from the National Society of Genetic Counselors and the American Civil Liberties Union, then, were classified as coming from "companies," in the same group as Myriad Genetics. This categorization demonstrated how the PTO understood patents, its publics, and its role in technology policymaking. By comparison, a 2015 FDA notice of a public workshop on electronic cigarettes and public health allowed the public to choose among twenty-eight categories, ranging from "Health Care Organization" to "International Public Citizen," when submitting its comments.[35] But at the PTO, consumers, or citizens more broadly, were not relevant constituencies; research institutions, companies, and lawyers would represent the interests of citizens by representing the interests of innovation and the market. This categorization also focused attention on the parties involved in producing innovation.

In September 2015, long after the Supreme Court case that I describe in the next paragraphs had concluded, the PTO finally released its report to Congress.[36] It concluded that there was very limited evidence that patents or licenses influenced innovation or health care, discounting a healthy body of scholarly work that had emerged by this time. It urged caution for any interventions in the patent system, echoing the position ultimately adopted by the California Institute of Regenerative Medicine's Intellectual Property Task Force that I discussed in the last chapter. In addition, the PTO sought to change the focus of the concern by urging greater attention to "the role of cost and insurance." These conclusions and recommendations reinforced the PTO's definition of patents and the patent system as limited in their scope. Because it did not see a relationship between patents and health care, it saw limited evidence relevant to its study. Instead, it offered evidence that insurance costs were hurting testing availability, reinforcing its understanding that access to genetic testing was a problem of health service provision rather than patents.

Packaging Access Concerns for the Courts

As advisory committees, the PTO, and Congress slowly considered the possibility that patents might hurt access to genetic medicine, civil society groups took matters into their own hands. Knowing that the courts could produce the most decisive action, given the history and context of the US patent system, they launched a legal case that would ultimately force the issues onto the national stage. But in order to accomplish this, they had to not only transform their frustrations but also themselves so that they could conform to the rules of the courtroom and the patent system more generally.

In 2006, Tania Simoncelli, a science and technology policy specialist at the American Civil Liberties Union, learned both about human gene patents and the scholarly concerns about their impacts on research and access to health care. She was particularly frustrated by the BRCA gene patents and Myriad's aggressive efforts to defend its monopoly, because of her own family history of breast cancer. [37] She brought the issue to Chris Hansen, a senior ACLU attorney, and introduced the idea of challenging human gene patents by suing Myriad and the PTO.[38] Over the next couple of years, she, Hansen, and other ACLU attorneys explored the possibility of organizing a court case that focused on Myriad's BRCA patents but challenged human gene patents generally. They received advice from law professors, social scientists, genetic counselors and medical geneticists, pathologists, and physicians.[39] They also conferred with individuals and organizations who had been affected by Myriad's patent strategy and gene patents more generally, who could serve as potential plaintiffs. Soon, PubPat's Daniel Ravicher joined the team, as he too had been investigating whether gene patents—and Myriad's patents specifically—might allow for a good legal challenge that could spotlight the problems that the current patent system posed for the public interest.[40]

This initial list of confidantes and experts gives us a sense of the motivations involved. If it had been a traditional patent case, the ACLU and PubPat might have focused on gathering insights from patent lawyers and scientists who could advise them on the patentability of the BRCA genes on technical and legal grounds. But they were driven by concerns about the socioeconomic, health, and moral implications of these patents, which led them to enroll different allies in their proposed litigation.

Of course, the fact that the ACLU was involved at all suggested that the case was driven by nontraditional concerns. While it was America's largest and most powerful civil liberties organization and had a long history of using the legal system to achieve major victories in the name of the public interest,[41] the ACLU was new to the patent domain. In the past, it had only intervened in health care technology policy as it related to privacy issues. Simoncelli's position as science advisor was, in fact, completely new.[42] But the ACLU decided to invest in the case, both because its staff saw the issues in terms of "civil liberties" (rights to research and health) and because the case felt "novel and interesting and intellectually challenging to lots of people in the organization."[43] Indeed, the BRCA patents seemed to create the perfect circumstances to question the social benefits of patents overall. Both the BRCA genes and BRCA gene testing were well-known and were connected to a common disease, breast cancer, that had motivated robust patient advocacy and support communities as well as public concern.[44] These patient communities might be willing to help, and at the very least might make the public-relations battle easier. Because Myriad's patent strategy had already produced extensive policy discussion, some in the scientific and health care communities might now be willing to make their voices heard. It would be yet another challenge—this time by a particularly high-profile organization that had the expertise and resources to launch both legal and public-relations campaigns—to how the US understood the role of patents as well as how the patent system had defined itself.

Achieving success would be extremely difficult. The ACLU and PubPat had a long legacy of advocacy failure to contend with. Previous legal interventions in the patent system had not only been unsuccessful, but had been essentially thrown out of US courts. And even though there had been some policy discussion about the impacts of gene patents on research and health care, it had generated a cascade of advisory reports without any policy change. This legacy meant that potential plaintiffs and supporters might initially be reluctant to participate and that courts might be unsympathetic. In addition, while Ravicher brought patent-law expertise and an emerging record of challenging the patent system in creative ways, he had no record of engaging in this kind of highly visible and complex legal fight.

Informed by this legacy, the ACLU and PubPat packaged their concerns in ways that would ultimately lead them to the Supreme Court. Well aware of previous jurisprudence that had restricted standing requirements, including the failure of the ALDF lawsuit against animal patents in the early 1990s, their first challenge was to choose a list of plaintiffs that would stand up to legal scrutiny. This would be quite difficult initially. While many supported the effort, taking on Myriad and the patent system was seen as too risky and quixotic, and potentially damaging to a plaintiff's reputation.[45] Geneticists, pathologists who ran laboratories, and genetic counselors worried that it might hurt their ability to negotiate compromises with the universities and companies holding gene patents.[46]

However, given the constraints of US legal-standing rules that require plaintiffs to have some sort of injury, the ACLU and PubPat couldn't challenge the patents themselves. So, they slowly and carefully assembled plaintiffs who all claimed some sort of injury resulting from Myriad's patent. Myriad had stopped plaintiff Arupa Ganguly, a geneticist at the University of Pennsylvania, from conducting research and clinical practice related to the BRCA genes. So the original complaint, launched at New York's federal district court, stated, "Dr. Ganguly, the co-director of the laboratory was ready, willing, and able to resume research and clinical practice if the patents had been invalidated. If they are invalidated now, she would seriously consider resuming clinical practice that is now prohibited."[47] Individual women and breast cancer organizations also joined the lawsuit, arguing that Myriad's patent monopoly had prevented them from having confirmatory laboratory testing done; because they could not access a second opinion, they had poorer health care. Plaintiff Breast Cancer Action, a patient advocacy group, stated when the lawsuit was filed that "Breast Cancer Action is deeply concerned by the barriers to research and clinical care created by Myriad's patents. They endanger our members' health and threaten the progress of patient-based health care. We believe it is wrong for the government to give one company the power to dictate all scientific and medical uses of genes that [we all have] in our bodies."[48] By gathering plaintiffs with different kinds of complaints against Myriad, the lawyers hoped to both gather diverse support across communities and preemptively offer a variety of answers to concerns regarding legal standing, and thus avoid challenges on that basis.

In total, ACLU and PubPat lawyers gathered twenty plaintiffs. The lead plaintiff was the Association for Molecular Pathology, which represented scientists who developed and conducted diagnostic testing, but the list included individual medical geneticists, pathologists, molecular geneticists, genetic counselors, and women both living with and at risk for breast cancer. It also included organizations focused on medical genetics, breast cancer patient advocacy, and women's health. Taken together, the plaintiffs embodied the access and control concerns that drove the case. They looked very different from the usual participants in the patent system, challenging the chasm between patent law and other areas of law (particularly those wrestling with policy issues). An ACLU representative reflected,

> We put this case together as a civil rights case, not as a patent case. That's part of why the Patent Bar [lawyers admitted to practice in the patent system] is so uncomfortable with the case. Our plaintiffs are civil rights plaintiffs, not patent law plaintiffs, which is why we're having so much standing problems. It's one of the half a dozen ways in which this case is really novel for the patent world. But, I will say, totally garden variety when it comes to the civil rights world. You couldn't be a more typical civil rights case than this. *It's just that the Patent Bar doesn't understand the fact that civil rights apply to patents* [emphasis added].[49]

While framing the case in civil rights terms may not have gained much support among the Patent Bar, which would likely have opposed the case regardless of its focus, it did help the plaintiffs achieve acceptance among the scientific, medical, and patient advocacy communities. This allowed them to accomplish their second strategic move: to build a large group of official supporters. Eventually, the group would include Nobel Prize–winning British scientist John Sulston, the American Medical Association, and the National Women's Health Network. Some filed expert declarations, while others wrote official statements of support. [50] These declarations and statements provided evidence of how gene patents, and the BRCA patents in particular, had infringed upon rights to research and access to health care.[51] Some also questioned whether Myriad deserved the patents at all, calling attention to the international network of researchers involved and the major breakthroughs along the way credited to other researchers.[52] Lawyers hoped that enrolling these allies would bring further support for and attention to the case. These participants would galvanize their social networks, and because the vast majority of them had established power and credibility in other areas of public policy, they would likely be able to garner media and public sympathy.

The final strategic decision was to expand the lawsuit's focus beyond distributional concerns to argue that human gene patents violated US patent law's "product of nature" doctrine. There was little question that the plaintiffs and their supporters, as well as the lawyers, were primarily concerned about the impacts of these patents on research and health care access. After all, the official complaint's introduction stated,

> Because of the patents and because Myriad chooses not to license the patents broadly, women who fear they may be at an increased risk of breast and/or ovarian cancer are barred from having anyone look at their BRCA1 and BRCA2 genes or interpret them except for the patent holder. Women are thereby prevented from obtaining information about their health risks from anyone other than the patent holder, whether as an initial matter or to obtain a second opinion. The patents also prevent doctors or laboratories from independently offering testing to their patents, externally validating the test, or working cooperatively to improve testing. Many women at risk cannot even be tested because they are uninsured and/or cannot afford the test offered by Myriad.[53]

But the lawyers knew that the courts would be unlikely to adjudicate these concerns. Unlike Europe, where the EPO and other governing institutions in the patent system had at least shown a willingness to engage with moral concerns via the *ordre public* clause, the United States had no such record. In fact, by then, the US patent system had demonstrated itself to be impervious to the long list of moral, social, economic, and environmental critiques that had emerged since the 1980s. And while distributive concerns were getting increasing policy attention in the United States, they had generated no change to date. Furthermore, such issues might be seen as "policy" questions to be considered by the executive and legislative branches rather than matters of "law" to be considered by the courts. There was even a chance that focusing on these issues might get the case thrown out of court, because they were so strange in the context of the patent system's customary considerations.

Despite the concerns that drove them, therefore, the ACLU and PubPat argued that their "cause of action" was a violation of the "product of nature" doctrine. "Because human genes are products of nature, laws of nature and/ or natural phenomena, and abstract ideas or basic knowledge or thought, the challenged claims are invalid under Article 1, section 8, clause 8 of the United States Constitution."[54] In sum, they designed a lawsuit that had the greatest chance of success given the narrow opening revealed by the previous decades of activism surrounding the US patent system. This also risked a decision that would not address their concerns directly. But they calculated that given the political history of life-form patents over the previous decades, this would give them the best chance of success.

These strategies paid off, at least partially. In March 2010, the federal district court accepted the legal standing of all plaintiffs and issued a summary judgment in their favor.[55] In his ruling, Judge Robert Sweet summarized the evidence presented by the plaintiffs, Myriad, and the additional briefs, regarding whether gene patents hurt innovation and health care. But the summary judgment process required the court to focus on matters of "law," rather than "fact." Therefore, Judge Sweet focused on interpreting the patent statutes: he decided that genes were "products of nature" and therefore unpatentable.

This was a momentous event. Many observers pointed to the novelty of the court's decision: for the first time since *Chakrabarty*, the US courts had limited the patentability of life forms.[56] But it was also important because it took the distributional concerns, which had originally motivated the plaintiffs and captured the public's interest, off the table and focused the case solely on the "product of nature" doctrine. In order to achieve success, the ACLU and PubPat had had to sacrifice the concerns that motivated them.

Until then, few decision makers or stakeholders in the patent system had taken the *AMP et al. v. Myriad* lawsuit seriously. But things changed when Myriad appealed the District Court's decision to the Court of Appeals for the Federal Circuit (CAFC). Before the CAFC heard oral arguments, it asked President Barack Obama's Department of Justice (DOJ) to weigh in. Although the United States government, through the PTO, had granted gene patents for decades, the DOJ reevaluated the government's position and filed a brief stating that isolated genes should not be patentable because they were "products of nature." (In its judgment, cDNA could be patented because it had been partially manipulated: the nonfunctioning parts of the genes had been removed.)[57] The solicitor general then argued the case in person, alongside the ACLU lawyers, at the CAFC. This was quite a shift. The DOJ broke with an agency in its own executive branch. Furthermore, the case was deemed so important to the Obama administration that it defended its interpretation vigorously and publicly, and it deployed its DOJ and solicitor general to perform these tasks.[58] All of this departed from the customary protocol that would have left such intervention to the PTO.

In summer 2011, a three-judge panel of the CAFC, which had become well known for supporting the expansion of patentable subject matter,[59] asserted the patent system's traditional approach and decided (2–1) in favor of Myriad. With this decision, it reinforced the definition of patents as legal and technical objects that had stimulated the growth of a pivotal industry. And, as we have come to expect by now, with this also came narrow definitions of who had the legal standing to participate in the system.

Judge Alan Lourie, who trained initially as a chemist before becoming a patent lawyer, wrote the majority opinion. On the legal question of patentability, he argued that because the patented molecules containing the BRCA genes were "markedly different" from genes inside the body, they were not "products of nature."[60] He also denied legal standing for all but one of the plaintiffs. Only medical geneticist Dr. Harry Ostrer, he argued, could claim clear "real" and "immediate" injury "caused" by the BRCA1 gene patent that could be redressed, because Myriad had stopped him from performing BRCA testing and because he claimed that he would immediately begin testing again if the patents were revoked.[61] The court concluded that all of the other plaintiffs, despite their various claims on the basis of scientific, health, and even economic harm, could either not claim direct injury or could not demonstrate that patent revocation would redress their injury. By defining legal standing in this way, the court continued to limit the plaintiffs in patent cases to a limited group of economic actors whose market position had been hurt directly by the patent holder. For example, because Myriad was not the sole cause of their health injury, women concerned about their risk of breast cancer could not sue. But this approach had broad impact beyond the Myriad patents: the courts would be unlikely to see patent cases that raised moral or socioeconomic issues and thus would be unable to weigh in on patents that raised the most public concern. In addition, the courts would be unable to weigh in on the growing set of challenges that questioned the distributional effects of the patent system overall. These limitations were particularly significant because in the United States, patents were defined primarily in legal terms, and the court system was seen as the central and ultimate authority for patent decision making.

In her concurrent opinion, Judge Kimberley Ann Moore, who had a background in electrical engineering as well as patent law, disputed one of the plaintiffs' arguments, specifically the idea that gene patents stifled innovation: "The biotechnology industry is among our most innovative, and isolated gene patents, including the patents in suit, have existed for decades with no evidence of ill effects on innovation."[62] She supported this statement by quoting an article from the *Texas Law Review* that summarized statistical work on the effects of biotechnology patents on innovation and concluded that there were "few clear signs" that they were "adversely affecting biomedical innovation."[63] To tinker with the PTO's existing practices and the "settled expectations" of the biotechnology industry by disallowing gene patents, she argued, could well hurt rather than promote innovation. Given these factors, as well as the approach to patent oversight described in the Constitution and patent statutes, and Congress's silence on issues related to patents on life forms, Moore suggested that any decision to prohibit gene patents (particularly on the basis of innovation concerns) should come from Congress. Overall, however, she echoed the claims of the past, emphasizing that the US patent system's role was minimal and simply created the conditions for innovation and economic growth. To interfere further, she suggested, would unsettle the experiential knowledge of market players and create intolerable and unnecessary uncertainties. To her, the evidence was not strong enough to take such a decisive step. Although Moore's opinion did not have the force of law, it gives us a fuller sense of the majority's position and worldview.

It is worth noting Moore's great emphasis on settled expectations. This echoed the focus on certainty that traditional organized interests had articulated in previous debates. To them, to consider distributional implications would hurt the patent system's predictability. The current system operated according to seemingly objective procedures, treating all patent applications the same, which prospective applicants could anticipate and therefore trust. But a system that considered impacts would jeopardize this predictability. They argued that this would make inventors less likely to participate and eventually hurt innovation and the market overall.

The ACLU and PubPat appealed to the US Supreme Court in 2012. But after ruling in a related case, *Mayo v. Prometheus*, that in order to receive a patent, inventors must do more than "recite a law of nature," the Supreme Court sent the *AMP v. Myriad* case back to the CAFC. The CAFC essentially repeated its first decision,[64] and finally, in April 2013, the Supreme Court heard the AMP case. The discussion focused, as it had in the lower courts, on whether isolated genes were products of nature. Policy concerns loomed in the background but the lawyers and justices focused on stimulating innovation, rather than access to health care. In his opening sentence Solicitor General Donald Verrilli, arguing on behalf of the Obama administration (for neither party), stated, "Enforcing the distinction between human invention and a product of nature preserves a necessary balance in the patent system between encouraging individual inventors and keeping the basic building blocks of innovation free for all to use."[65] And later in the hearing, Justice Elena Kagan asked the ACLU's lawyer the appropriate way to decide how much incentive would be enough for innovators.[66] Distributional concerns, however, were only a few feet away. On the steps of the Supreme Court, dozens of breast cancer activists gathered in protest, making clear that they saw a link between patentability of the BRCA genes and access to quality BRCA gene testing (figure 5.1).

In June 2013, the Supreme Court ruled unanimously that isolated human genes were unpatentable because they were "products of nature."[67] Policy concerns were virtually absent from the ruling, which was based on patent-law statutes and case precedent. The court left open the possibility that cloned DNA (cDNA) could be patented, because the process of removing exons from human genes involved sufficient manipulation to allow patentability. There was some question of whether cDNA would fulfill the PTO's "nonobviousness" standard, but this decision created a glimmer of hope for the biotechnology industry.

US case law had turned on questions of "nature." As a result, the knowledge brought to bear on the issue was either legal or technical. Indeed, in addition to patent-law statutes and case precedent, during the hearing the Supreme Court took special note of an amicus brief filed by Eric Lander, a geneticist and the founding director of the Broad Institute of MIT and Harvard, which argued that cDNA strands could be found in nature.[68] These issues had never arisen in similar European discussions. Rather, debates over hESCs, and genes, as we will see below, which could have raised questions regarding the distinctions between nature and technology or between discovery and invention, had instead focused on questions of "life" and *ordre public*. In the European hESC patent cases described in the previous chapter, for example, the EPO and ECJ focused on the destruction of human embryos and the ethics of producing markets in living things. Consequently, as we have seen, the European patent system had begun to value different kinds of experts. But the focus on nature in the United States privileged scientific and legal expertise. The AMP case, and

Figure 5.1.
Protester challenging the patentability of human genes in front of the US Supreme Court on the day it held hearings on the issue, in April 2013. Patient and environmental groups organized the protest.

the resolution of the human gene patent issue generally, also highlight the difficulty of forcing the US patent system to consider distributional concerns. Despite the policy attention to the issues and the engagement from powerful groups like the ACLU, it was virtually impossible to produce action that addressed these frustrations directly. The United States had long rejected the idea that patents had distributional effects, and this rejection was institutionalized in its legalistic nature, its standing rules, its focus on technical and experiential expertise, and its discomfort with broad public participation. As a result, the ACLU, PubPat, and the plaintiffs they represented had to thread the eye of a narrow legal needle that could address their immediate concerns only indirectly.

Patents and the Right to Food

The pitfalls of this indirect approach to distributional concerns would be made clear in PubPat's next major court case. Buoyed by his success with *AMP* and his new visibility as a public-interest patent lawyer, Ravicher, on behalf of ninety-five agriculture and food-safety membership organizations, seed businesses, farms, and farmers, challenged seed patents owned by giant agricultural company Monsanto. He argued that Monsanto's enforcement of its patent rights, which had included infringement lawsuits against many farmers, was harming both farmers and consumers. The Organic Seed Growers and Trade Association, OSGATA, which represented the organic farming industry, was the lead plaintiff.

Ravicher had very little room to maneuver. The courts had never before taken such concerns seriously in the context of the patent system. But he and the plaintiffs adopted an ambitious strategy. Rather than taking a more conventional approach and making his case on the basis of the product of nature doctrine or other law that had a history of consideration by the courts, they argued that Monsanto's patents were invalid due to the "moral utility" doctrine. As I discussed previously, this doctrine had made only rare appearances in the history of patent law and in the history of life-form patents in particular. This approach echoed the emerging strategy of life-form patent challengers in Europe. Ravicher and the plaintiffs in this case, like Kein Patent auf Leben and its allies, were no longer content to eliminate the patents on any grounds. After the success of the *Myriad* lawsuit, they wanted the patent system to explicitly consider moral and socioeconomic concerns. Ravicher acknowledged his exceptional approach but argued that the "moral utility" argument was "the closest to the substantive concerns the clients have with this [patent]."[69] The complaint stated, "Because transgenic seed, and in particular Monsanto's transgenic seed, is 'injurious to the well-being, good policy, or sound morals of society' and threatens to 'poison people,' Monsanto's transgenic seed patents are all invalid."[70] It specified "the perils of transgenic seed" in terms of economic harms and health and environmental risks, and also stated that the technology had not achieved its promised benefits to increase crop yield, lower farmers' costs, and decrease pesticide use. The complaint also invoked rights-based arguments that had socioeconomic consequences. It emphasized that the patents made it impossible to engage in activities that farmers "have the right to pursue (e.g., growing crops they wish to grow on their land)" and that the increased concentration caused by the GM seed patents "has diminished consumer choice and slowed innovation."[71] In other words, these patents hurt society by placing control over an invention into a single owner's hands, and they specifically infringed upon the rights and freedoms of farmers, innovators, and consumers. The complaint was, of course, linking patents to their use in the world.

The District Court's response illustrated the courts' limitations. In February 2012, Judge Naomi Buchwald dismissed what had become known as "the OSGATA case," ruling that the plaintiffs had no legal standing to sue. There was no "grievance" or clear market interference, she concluded, and therefore no "case or controversy" at stake.[72] Monsanto had neither sued nor threatened to sue the plaintiffs. While plaintiffs had argued that there was a real fear of future harm, because between 1997 and 2010 Monsanto had filed 144 patent infringement lawsuits against US farmers and threatened to sue countless others, Judge Buchwald responded that this evidence demonstrated a low threat in the context of the two million US farms.[73] In other words, the judge and the challengers viewed the evidence differently based on their positions. This echoed the earlier disputes over whether patents stifled research. To the researchers and to the farmers involved, the few cases were too many. But to this court, which viewed the question through the lens of legal standing, they were not enough.

Finally, Judge Buchwald derided plaintiffs' efforts to use the court to address moral and regulatory issues. Plaintiffs were, in her opinion, attempting "to create a controversy where none exists."[74] Like dozens of patent-system insiders and decision makers before her, Judge Buchwald argued that the courtroom was not the place for citizens to air their grievances regarding the implications of patented seed for organic farmers, or the implications of genetically engineered crops more generally. She even observed that "plaintiffs' argument is baseless and their tactics not to be tolerated."[75] Furthermore, the court could not comprehend or act on patents' distributional effects because its legal-standing requirements prevented it from doing so. Even though farmers were, at first glance, active participants in the same agricultural marketplace as the patent holders, the courts could not consider their concerns, because there was no active "case or controversy." This case, coupled with the *AMP v. Myriad* victory, demonstrated the challenges faced by citizens concerned with the distributional effects of patents in the United States. They had begun to focus their attention on the courts because other institutions involved had been unable or unwilling to take significant action, and other policy domains were not stepping in to address broader worries that technologies exacerbated economic inequalities under some circumstances. But the courts constrained these actors and their concerns in ways that not only made their goals extremely difficult to achieve but also distorted them, even when they achieved some success.

Transforming the Public Sphere
While US patent-system institutions were reluctant, and often simply lacked the capacity, to consider the idea that patents had distributive implications, there were some signs that these cases were shifting the public mood. In contrast to the days after the *Chakrabarty* decision, when the *New York Times* emphasized the limited nature of the patent system, news outlets now seemed to regularly run stories about the implications of life-form patents on research, agriculture, and average citizens.[76] During the *Myriad* case, for example, stimulated in part by the ACLU's public relations machine, there were hundreds of stories that showcased worries about the impact of gene patents on health care access and quality.[77] One of the plaintiffs, genetic counselor Ellen Matloff, described in women's health magazine *Self* how Myriad had refused to incorporate insights from new research into its testing protocol, which meant that women received incorrect information about their BRCA gene-mutation status. She recalled an episode in which she contacted Myriad after scientists at the University of Washington had discovered that the company's test missed about 12 percent of disease-causing mutations in breast cancer patients. She asked Myriad if her lab could offer patients that additional analysis, since it would be easy for them to do so: "They said no, that *they* were going to offer the test once they'd completed their research. More than a year passed before they started offering it, now called the BRACAnalysis Rearrangement Test [BART]. We had to sit here that whole time knowing some patients had mutations that were being missed [emphasis in original]."[78]

Meanwhile, movie studios released many documentaries on these topics.[79] One of the most widely viewed, Academy Award-nominated *Food, Inc.*, interviewed farmers who had been the subject of Monsanto's patent-infringement lawsuits.[80] Reputable newspapers and magazines also published related editorials. Questioning the socioeconomic harms of patents, an op-ed in the investment-focused website *The Daily Finance* linked patents to "rising food prices."[81] *The Economist*'s editorial board wrote an article concluding that gene patents were simply "bad" for disease prevention.[82] Movie star Angelina Jolie's widely read op-ed in the *New York Times*, which focused on her decision to have prophylactic breast and ovary removal surgeries after learning of her BRCA mutation status, referred to the "obstacle" created by Myriad Genetics's testing monopoly.[83] In the title track to his 2015 album *The Monsanto Years*, Neil Young sings, "The farmer knows he's got to grow what he can sell, Monsanto, Monsanto / So he signs a deal for GMOs that makes life hell with Monsanto, Monsanto / Every year he buys the patented seeds / Poison-ready they're what the corporation needs, Monsanto . . ."[84] Finally, surveys showed that a majority of citizens saw gene patents as damaging health care and research.[85] The media and the public had clearly begun to see patents as potentially having negative socioeconomic implications, and they had become concerned. This perspective was increasingly in tension with the presumptions within the US patent system, which continued to hold fast to the idea that patents should be distinguished from the effects of monopolies created by patent holders.

EUROPE

European countries had long seen patents as having distributive implications, and they also saw a role for the patent system in shaping the socioeconomic impacts of technology. After all, as I discussed in chapter 1, many European countries had banned patents on food and pharmaceuticals in the nineteenth century, because such patents might interfere with the public's basic needs. But this focus had waned over time. Food and drugs are now patentable throughout Europe. And while the European Union's Biotech Patent Directive had generated intense debate about the social and economic implications of biotechnology patents for citizens in both the developed and developing world, the final legislation validated only a few of these concerns. Human genes were deemed patentable despite worries that such patents might produce monopolies that would increase costs for health systems and force citizens to rely on lower-quality genetic tests. The law also validated patents on genetically modified plants, which many worried would hurt both farmers and food availability. But it had given each farmer the privilege to save patented seed across generations, which, when coupled with the EPC's prohibition on patents covering plant and animal varieties, demonstrated some effort to maintain farmers' traditional practices.

Considering Distributive Concerns in Bureaucratic Decision Making
As the European Parliament scrutinized the BPD in the mid-1990s, Myriad Genetics applied for EPO patents on the BRCA genes. Simultaneously, the company demanded that European providers of BRCA testing shut down their services or risk patent infringement, expecting that the EPO would soon grant the patents.[86] In the United States, this approach had been successful initially.

But in Europe, scientists, physicians, public-health officials, and patients responded quickly in an overwhelmingly negative manner.[87] At first, they fought Myriad inside their own national health systems. Then they questioned the gene patentability proposed in the BPD by sending letters to and visiting members of the European Parliament, national government ministers, and the directors of medical research institutions, and by forming coalitions with environmentally focused groups, including Greenpeace and the Green Party.[88] They argued that gene patents would require public-health systems to pay high licensing fees and thereby increase costs for public-health systems. One patient advocate in the United Kingdom explained her perspective as follows: "If [gene patents are] accepted the upshot will be that our NHS service will have to pay a huge royalty to Myriad for every genetic test undertaken. . . . I would not like to see our NHS collapsing totally under the weight of royalties to biotech companies."[89] They also worried that these patents would force public-health systems to adopt inferior and inaccurate testing approaches. Myriad did not require specialized genetic counseling alongside BRCA testing, they pointed out, which meant that clients often received complicated risk information and had to interpret it on their own.[90] These efforts to oppose Myriad were unsuccessful, however, partly because they coalesced as the BPD debate was coming to a close. As I discussed in chapter 2, the final legislation allowed the EPO to issue patents on genes "isolated from the human body," taking a position that the US Supreme Court would eventually reject.[91]

Critics then turned their attention to the EPO, which granted Myriad three patents on the BRCA1 gene in 2001. Almost immediately, twelve organizations—including research institutes, hospitals, genetics clinics, political parties, government ministries, and Greenpeace—filed oppositions. Twenty patient advocacy groups, scientific societies, organizations representing health professionals, and health ministries offered official support. Like their US counterparts, almost all of them mobilized because of their concerns regarding how these patents would affect health care. (Probably because of the various exceptions for research or experimental use at the national and European levels, the effects of these patents on research did not seem to be a primary concern.) The Institut Curie, the French cancer research organization where scientists discovered that Myriad's test missed large genomic rearrangements, explained its opposition: "The Institut's initiative also aims to curtail the consequences in France and in Europe of a monopoly which will jeopardize the development of research, hinder access to testing, and furthermore go against our approach to public health, which is based on our committment [*sic*] to the comprehensive and multidisciplinary care of high risk patients."[92]

Most opponents used the *ordre public* clause as the vehicle to make their case. Some suggested, echoing the ACLU's concern in the United States, that by turning genes into commodities, the EPO limited an individual's right to access vital health information about herself. Such gene patents were therefore unethical. The representative of the Belgian genetics societies noted that

> patents are provided with safeguards, . . . which limit them to certain embodiments and not to any form of implementation [of] the invention. In the field of patenting genes there is the possibility of obtaining a monopoly position when the patent right piggy backs on the unique nature of genetic material, e.g., of the human genome. . . . The patent rights granted to Myriad allow Myriad to prevent all diagnosis using mutation analysis of the human genome. . . . This is unethical. . . . A similar argument is that genetic information is a very unique and personal description. Each person in the member states should be able to have this inherited information read and interpreted by the organisation of its choice. This intensely personal information must be kept patent free. The opposed patent can prevent this free access to personal information. We consider this unethical."[93]

Greenpeace echoed this sentiment, noting broadly that gene patents restricted access to genetic resources. This was particularly problematic in the BRCA case, it argued, because Myriad had a monopoly over DNA that affected one group of patients who were at risk for a life-threatening disease, and these patients had expressed opposition to the patents.[94]

But they also defined *ordre public* in socioeconomic terms, echoing the clause's early meaning. Some opponents provided evidence that Myriad's patents violated *ordre public* by citing recent decisions by international governing bodies to recognize moral limitations on intellectual property rights, particularly in the context of health. Italy's Materdomani Hospital explained that the World Trade Organization's Trade-Related Aspects of Intellectual Property (TRIPS) Agreement, signed in 1994, provided exceptions on the basis of "public order, morality, and the protection of human life and health."[95] The Swiss Social Democrat Party referred to a resolution of the UN High Commission on Human Rights that asked parties to the International Covenant on Economic, Social, and Cultural Rights (which included many European countries) to ensure that current intellectual property rights regimes would protect "the fundamental nature and indivisibility of all human rights, including the right of everyone to enjoy the benefits of scientific progress and its applications, the right to health, the right to food and the right to self-determination."[96] These legal frameworks, opponents argued, demonstrated why the *ordre pub-*

lic clause should be understood in terms of access to health. And if it was interpreted in this way, they suggested further, Myriad's patents demonstrated clear violations.

Some of them also attacked Myriad's patents on grounds they predicted might be more successful. Just like their US counterparts, who had largely focused on "product of nature" issues inside the courtroom, European opponents questioned the BRCA patents' novelty, the degree to which they represented an inventive step, and whether the invention had been sufficiently disclosed.[97] Lawyers for the Belgian Society of Human Genetics also tried another strategy, focusing on EPC language prohibiting patents on therapeutic or "diagnostic methods practiced on the human or animal body." As discussed in chapter 1, the EPC's authors had included this text to ensure that patents did not interfere with medical care. It had rarely been tested,[98] but the Belgian Society saw it as another opportunity to force the EPO to consider the public health and social impacts of patents. Previous case law had invalidated patents on diagnostic or therapeutic methods, they observed, because they had required a "doctor step." "The purpose of Art. 52(4) EPC is to exclude any patent which seeks to intercede between the patient and his or her medical adviser. . . . The diagnosis of a hereditary disease, for which there is no cure, and which is linked to a painful death and whose traumatic effect has been and still is highly publicised in the press has to be treated with care by medical practitioners."[99] In order for the BRCA gene to have any meaning, they suggested, it had to be part of a diagnostic and therapeutic regimen that involved a medical practitioner (it had to be interpreted, for example, in the context of family history, lifestyle, future goals, and other risks). Because a patent on the BRCA genes would interfere with the medical practitioner's work, they argued, it was invalid.

Soon after the official opposition process began, the European Parliament amplified critics' concerns. It issued a nonbinding resolution declaring that gene patents violated the *ordre public* clause in two ways. (A show of hands decided the final vote, so no count was recorded. According to the EP's Citizens' Enquiries Unit, this "is a usual procedure when the item is considered uncontroversial across the political groups.")[100] The EP recalled moral concerns regarding commodification by arguing simply that European patents should "not violate the principle of non-patentability of humans, their genes or cells in their natural environment."[101] But it also invoked socioeconomic concerns, noting that Myriad's patents "could seriously impede or even completely prevent the further use of existing cheaper and more effective tests for the breast cancer genes BRCA1 and BRCA2; . . . could have an unacceptable detrimental effect on the women concerned and constitute a serious drain on the funds of public-health services; . . . [and] could seriously impede the development of and research into new methods of diagnosis."[102] While the BPD had recently allowed human gene patents, parliamentarians now seemed to argue that they were immoral in several ways. Furthermore, they tied the implications of patent-based monopolies to the patents themselves, a connection that US decision makers had been reluctant to make.

As it did this, the European Parliament envisioned a patent system that was responsible for these public-health implications and for responding to public concerns. In fact, it explicitly rebuked the EPO on this basis, demanding again "a review of the operations of the EPO to ensure that it becomes publicly accountable in the exercise of its duties."[103] This text was identical to language in the EP's resolution on human embryonic stem cell patents, issued just a few months before. To these EP members, the EPO was not performing its role or responsibilities appropriately because it was not adequately considering moral and distributional issues in its patent decisions. Although the US Congress had asked the PTO to review the impact of patents on genetic testing in its America Invents Act, it had never challenged the accountability of its patent bureaucracy. In fact, its treatment of the PTO suggested that it saw accountability in terms of procedural objectivity and the issuance of patents, rather than in terms of balancing between the interests of the inventor and the public.

Because the EPO was not part of the European Union, the European Parliament had no official jurisdiction over the bureaucracy. Furthermore, the resolution was nonbinding. Still, it was the second critical resolution passed with a majority vote among elected representatives of Europe on matters that had garnered a fair bit of public attention.[104] Therefore, its vision of patents and its challenge to the EPO would be hard for the bureaucracy to ignore.

The EPO held oral opposition proceedings in 2004 and 2005 on all three Myriad patents under challenge, with Myriad and the opponents present. Myriad engaged with opponents' challenges to the novelty, inventiveness, and industrial application of the invention, and to the description in the patent, but it dismissed the relevance of moral and distributional concerns, including the *ordre public* clause. Taking an even stronger stand than the University of Edinburgh had in the hESC patent case, it called the concerns "irrelevant and false,"[105] and charged that the claims represented "an abuse of the procedure and as far as [Myriad's lawyer] incurred costs for having to deal with these arguments, he requested apportionment of these."[106] To an American company and its lawyers, opponents' arguments very likely made no sense. Public order and morality, however they were defined, were simply separate subjects not to be discussed in the context of patentability.

The EPO eventually decided to revoke one of Myriad's BRCA patents and substantially narrow two others.[107] The bureaucracy justified these decisions by arguing that the invention was not novel, did not involve an inventive step, and was insufficiently described.

In its decision, the EPO agreed with Myriad that the *ordre public* clause was irrelevant. It argued that the clause should be understood solely in terms of "legal and ethical values": "[The clause] can only apply in rather exceptional cases, namely, where the publication or exploitation of the invention as claimed is in conflict with basic legal or ethical values."[108] If the invention itself or the process of its creation was morally offensive, as in the case of hESCs, then commodifying it through patents would also be immoral and therefore a violation of the *ordre public* clause. In the BRCA gene patent case, however, opponents were arguing that the patent itself was immoral, not the invention.

But as the EPO drew a distinction between patents on *inventions* whose publication or exploitation was contrary to public policy or morality (like hESCs), rather than on *patents* whose publication or exploitation was similarly problematic (like the BRCA genes), it actually accepted the idea that patents had distributive implications. In fact, it noted explicitly that it could not consider the economic impacts of patents in its decision making because "the nature of the consequences of the exploitation of a patent (which derive from the exclusionary nature of private property rights), are the same for all patents."[109] Unlike the PTO and other US patent-system institutions, the EPO saw all patents as having distributive implications. Thus, it argued, it could not specially attend to the public-health consequences of the BRCA gene patents without creating an uneven playing field for inventors in different technical fields.

The EPO also responded to the European Parliament's concerns directly in its decision. After all, this was the second time in a year that a majority of parliamentarians had expressed displeasure. In its response, the EPO offered a very different vision of its role and its democratic responsibilities: "The Office is not the legislature, which has to balance conflicting interests and lay down legal rules. The EPO is an administrative agency which applies and interprets the rules laid down by the legislature. . . . The EPO is publicly accountable. The EPO is supervised by the Administrative Council of the European Patent Organisation. The transparency of the procedures before the Office is exemplary. Patent applications are published 18 months after filing or after the priority date. Any person can then inspect the respective file."[110] Here, it seemed to characterize itself as quite similar to its US counterpart. Its job was limited, it argued, to interpreting the law and being transparent about its procedures. Furthermore, it argued that it *was* publicly accountable to a quasi-legislative body, the Administrative Council. The EPO also reminded the Parliament that it had authorized human gene patents in the BPD, and that legislative body needed to authorize any changes. Overall, the Myriad case seemed to demonstrate a European patent bureaucracy that had retreated, in the case of human gene patents, to a limited, techno-legal understanding of its role.

This position did not stop critics. Challengers soon focused on patents covering a new type of innovation: conventional plants bred using biotechnology (and related processes). They were no different from conventional plants, but biotechnology had improved breeding efficiency. One technique, marker-assisted breeding, used molecular markers to track the inheritance of and select for a particular gene (or genes). The plants produced included broccoli with enhanced anti-carcinogenic properties and tomatoes with reduced water content.[111] To many farmers, food security groups, international development organizations, and groups concerned about agricultural biotechnology, these patents would disrupt traditional farming practices and jeopardize farmers' livelihoods even more than genetically modified crops. Because the patents covered both the plants and the processes of making them, it would force farmers who had developed these crops using conventional breeding techniques to license these patents or go out of business. This could then reduce options for consumers who did not want to eat foods produced using biotechnology. These groups also worried that the patents would restrict biodiversity and therefore food security, and ultimately reduce innovation by sending traditional breeders and farmers out of the market.[112]

To fight these patents, challengers created a new coalition entitled No Patents on Seeds! (NPOS), with international development organizations from Norway, Germany, and Switzerland, groups representing rural farmers from France, Italy, and Spain, a Dutch organic food association, KPAL, Greenpeace, and German GeneWatch, a biotechnology watchdog group.[113] NPOS filed multiple oppositions at the EPO, intervened in ongoing appeals processes on a variety of cases, and demonstrated in front of the EPO (figure 5.2). In describing its opposition to the broccoli patent, for example, it argued that "the patent is in conflict with the interests of farmers, market gardeners, breeders and consumers who do not wish to become dependent on international corporate businesses such as Monsanto."[114] They framed their complaints in terms of balancing the interests of society and the patent holders.

Given the outcome in the Myriad case, these civil society groups knew that

Figure 5.2. Protests against patents related to plants bred using biotechnology, at the EPO. Photograph by Faik Heller, for Greenpeace (October 27, 2014).

the EPO would be unsympathetic to considering these socioeconomic concerns within the scope of the *ordre public* clause. Therefore they focused on language in the EPC that prohibited patents on both "plant and animal varieties" and "essentially biological processes," arguing that the EPO was essentially allowing patents on conventional breeding processes and their products.[115] Put bluntly, they argued that the EPO was allowing patents on nature, which clearly was an error. But just as civil society groups had done previously, NPOS and their supporters made their primary concerns clear as well. They submitted dozens of statements from farmers and breeders associations, environmental groups, and average citizens that explained their grounds for challenging the patents. NPOS's statement, which was signed by thousands of citizens from across Europe, stated in part, "Agriculture, livestock, and breeding must remain independent and must not be dominated by the patent claims of industry and research institutions. Only then can food production be organized according to the wishes of the consumer [translated from German]."[116] They seemed to hope that the EPO would consider its decision in the context of this public concern about these patents.

As they had in previous cases, EPO personnel acknowledged this scrutiny on several occasions as the plant patents made their way through the opposition and appeals processes, but they focused on the technical and legal questions at stake. Eventually, the EPO's Enlarged Board of Appeal was asked for its opinion on whether these processes or products were patentable. (It focused on two patents: one covering anti-carcinogenic broccoli and the other cover-ing dehydrated tomatoes.) It concluded that while it was "aware of the vari-ous ethical, social and economic aspects in the general debate," "the issues referred to it relate to questions of law."[117] It then decided that, while it would not allow patents on these processes, it would allow patents on products stem-ming from these processes.

Again, as it had in the BRCA patent case, the EPO had shied away from con-sidering socioeconomic issues in its opposition and appeals processes. It had already argued that such issues were distinct from moral considerations that could be addressed through the *ordre public* clause, and now it was suggesting that it simply could not include them in its legal decision making processes. The story could have ended there. The EPO, which had become the central space for discussions of patentability in Europe, as the courts had become in the United States, had chosen not to incorporate distributional concerns into its patent criteria. While the European Parliament was unhappy with this, thus far it had only issued nonbinding resolutions that seemed unlikely to provoke major legal change. The fact that the tale continued and that European patent institutions continued to try to address these issues marks the real difference between the US and Europe. Indeed, the EPO's legal decisions only tell us a small part of this story.

Distributing Responsibility across Patent System Institutions

Although the EPO had refused to incorporate socioeconomic concerns into its patentability criteria, it began to consider these issues in other aspects of its work. This was a logical next step in the efforts that I described in chapter 3 to make the bureaucracy more socially aware. It tackled these issues explicitly in 2004, as part of its Scenarios project, a large-scale, multi-year initiative to en-courage strategic thinking about the future direction of the EPO and the role of the patent system in global society. EPO President Alain Pompidou devel-oped it in order to "take a lead in ensuring that the system remains fit for [its] purpose in support of innovation, competitiveness, and economic growth for the benefit of the citizens of Europe."[118] With this broadly stated intent, we could imagine various approaches. If the United States had embarked upon such a study, it might have focused solely on macroeconomic concerns. But at the EPO, personnel started by interviewing one hundred scholars and stake-holders from all over the world, who were asked to predict and discuss the different social, political, and economic pressures that the office might face by the year 2025.[119] The variety of people chosen to be interviewed is telling, particularly when placed in a comparative perspective. They spoke to individuals from inventors' organizations and high-technology companies as well as patent lawyers. But they also interviewed Christoph Then, representatives of international development organizations, patient advocates, philosophers, and social scientists who had testified on behalf of opponents in the life-form patent hearings, as well as high-profile critics of the modern patent system, including Indian environmental activist Vandana Shiva.[120] Even if the PTO had engaged in a similar project, its interview list certainly would have looked different. But through this process, the EPO certified the relevance of different kinds of participants and experts in the patent system, who by now were no longer so new. It also provided these players—including those who had ethical, social, and economic concerns about life-form patents—with an opportunity to shape the future of the organization.

The EPO's team used this interview data to construct four possible future scenarios. Although all of them touched on the concerns raised by the life-form patent challenges in some way, the "Trees of Knowledge" scenario addressed ethical and socioeconomic concerns directly and with an approach that would be unfathomable in the United States. In this scenario, the EPO's personnel tried to predict the kinds of dilemmas that might arise if the grassroots movements continued to grow, and public trust in government, industry, science, and technology continued to erode. In particular, authors of this scenario organized the discussion around three key questions: "How can public and private interest in IP [intellectual property] be reconciled for the benefit of society? How are the ethical and moral dilemmas raised by technology reflected by the patent system? Where should the limits of patentability be drawn? By whom?"[121] The fact that the EPO articulated these questions was significant indeed and gives us a sense of how it was starting to interpret its responsibilities.

In addressing these questions, the EPO acknowledged the growing pressure on the patent system coming from farmers, ecologists, consumers, and patients regarding the "economic and social consequences" of patents.[122] And it argued that the question of *whether* the patent system was responsible for these implications was irrelevant. It noted in the case of access to essential medicines, "Whether or not the patenting system is at fault for a failure to get drugs to those who need them is immaterial. Blame is laid at the door of the IP system by many forces in society."[123] Put differently, the EPO was saying that it didn't matter whether or not the patent system had these distributional effects. The public believed it did, and therefore the patent system and the EPO had to believe it and act accordingly.

Although the *Scenarios for the Future* report did not conclude with any policy recommendations, EPO administrators clearly realized that they had to respond in some way to the public's perception of the distributional implications of patents. It wasn't a hidden, technical, bureaucracy concerned with interpreting the law and serving patent lawyers, agents, and innovators. Rather, it was concerned with the role of the patent system in governing technology in order to achieve broad public benefit. Its constituency was the European citizenry, and it was taking a much broader array of knowledge and expertise into account.

EPO officials unveiled the report in April 2007 at a large event in its Munich office.[124] Major figures in European politics and policy—including German Chancellor Angela Merkel—participated, and the event was well-attended by the EPO's employees.[125] All EPO personnel received copies of both the final report and a book of the interview transcripts. Like the SeCa system and other efforts to transform the EPO's culture, the *Scenarios* report was intended to expose personnel—including examiners—to alternative types of knowledge and expertise in the patent system, and also to make them aware of their responsibilities to a much wider set of constituents.

The EPO then tried to address distributive concerns indirectly through a "Raising the Bar" initiative introduced the same year. EPO's president at the time, Alison Brimelow, framed it as an effort to improve the "quality" of the patents issued by the EPO.[126] In her mind, the European patent system had a serious problem. For years it had been recognized for issuing clear and narrow patents that could stand up against legal challenge, but now their quality had begun to decline. She argued that the EPO was now granting patents that were too broad and vague, which was hurting the innovation system. It was the EPO's responsibility to rectify this situation and "grant patents only for innovations with sufficient inventive merit meeting the needs of society."[127] This initiative encouraged a systematic approach to ensuring patent "quality" in the process of searching for prior art and patent examination. This included a greater focus on clarity and conciseness in the patent application, and on revising examination practices to ensure that they were more in line with the EPC rules on sufficient disclosure and inventive step.

While this initiative seemed to focus on "quality," it would also limit the power of patent-based monopolies. Indeed, these concerns, and the civil society challenges more generally, had clearly shaped the Raising the Bar initiative. In 2010, EPO's new president, Benoît Battistelli, articulated the connection between the patent-quality effort and the life-form patent controversies:

The EPO is very aware of its responsibilities to society in this field, and it applies the most exacting standards to its examination of applications for biotechnology patents. . . . This stringency means that only 28% of all biotechnology applications received by the EPO are granted, compared with an average of 42% in other technical fields. The legislation we apply in our examination procedure has been adopted by all the member states of the European Patent Organisation, including Germany. As a result of our strict examination procedure and associated legislation, the number of patents giving protection to plants and animals is declining.[128]

With narrower patents, patent-based monopolies would have less of an impact. Researchers could work around them, and negative socioeconomic consequences would diminish. In fact, intellectual property lawyers worried that the new initiative was essentially raising the standard for patentability.[129] More specific and technical patent applications might also make it more difficult for patent challengers to mobilize the public in opposition. Opponents had been successful in building campaigns partially because patents covered categories of life forms (e.g., genes, stem cells, plants, animals) that already had social meaning. While the United States would develop an Enhanced Patent Quality Initiative a few years later, its approach was quite different.[130] It focused, not surprisingly, on reducing error and increasing predictability, rather than on narrower and less controversial patents.

In sum, while the EPO had rejected the consideration of distributional concerns in its patent-examination criteria, it did not dismiss the idea that patents had such effects. And while it could have argued that it had no responsibility in this area, it did not. In keeping with its increasing efforts to demonstrate its responsiveness to the European citizenry and to redefine itself as a policy organization, it engaged in numerous political, administrative, and examination-focused efforts. But, they were often indirect.

While the EPO adopted an often circuitous route to distributional issues, national and European legislators stepped in more forcefully. Responding to concerns that patents on human genes might hurt access to testing and therapy, many countries amended their compulsory licensing laws in the early 2000s. As I have noted previously, most European countries already had compulsory licensing laws that allowed them to step in if a patent holder was acting against the public interest (including public health). Given their experience with Myriad Genetics, however, and their worries about human gene patents generally, these governments decided that this was not enough. France and Belgium, whose geneticists, public-health officials, patient advocates, and hospitals had officially opposed Myriad's patents at the EPO, passed new laws that allowed the government to issue compulsory licenses on gene patents in the interest of public health.[131] Meanwhile, Switzerland permitted compulsory licensing of gene patents in the event of anticompetitive practices (although public-interest grounds were already part of its patent law).[132] In all of these cases, the accompanying documents designed to shape interpretation of the laws referred to gene patents and their potential to limit diagnostic testing.[133] Like the European Parliament, these national governments seemed to conclude that limited access and higher costs of diagnostic tests could be tied to gene patents, and that this was a serious enough problem that it warranted action from decision makers throughout the European patent system. While the United States had prohibited patents on isolated human genes completely, it had defined the problem as a techno-legal one. But in Europe, which defined the issue in distributional terms, compulsory licensing became the appropriate solution.

Legislators intervened in the case of plant patents as well. In 2012, No Patents on Seeds convinced three German members of the European Parliament from across the political spectrum to convene a public roundtable on plant and animal patents.[134] Witnesses included Then, Fritz Dolder (a lawyer who had helped Then with the plant-patent oppositions as well as other EPO challenges), representatives from farming and breeding organizations in Spain and the Netherlands, a German Catholic organization focused on international development, and the Mars food company's global programs initiative. They made the same arguments as they had in their EPO oppositions. In September 2012, NPOS presented to the president of the European Parliament a petition with signatures from almost three hundred organizations and seventy thousand individuals across Europe. The petition asked the EPO to revise European patent law: "In many cases these patents cover the whole chain of farm and food production. . . . Such patents create new dependencies for farmers, breeders, and food producers. This has to be regarded as misappropriation of basic resources in farm and food production and as general abuse of patent law."[135] To address these concerns, NPOS argued, the EP had to "support clear regulations that exclude from patentability plants and animals, genetic material and processes for breeding of plants and animals and food derived thereof."[136]

These arguments and concerns echoed those made by Ravicher's clients in the OSGATA case in the United States, but in Europe the outcome was quite different. EP members held hearings on the matter, and in May 2012, the EP passed a nonbinding resolution with 354 votes (192 against, with 22 abstentions) supporting the EPO's decision to prohibit patents on smart and conventional breeding techniques and asking it to exclude from patentability the products of these processes.[137] It also asked the European Commission to issue a report on the potential implications of patenting plant breeding for "the breeding industry, agriculture, the food industry and food security."[138] In making this request, it observed that plant breeding was "an essential prerequisite for the security of the food supply and, to some extent, of the energy supply," and that "excessively broad patent protection can hamper innovation and progress and become detrimental to small and medium breeders by blocking access to animal and plant genetic resources." For European parliamentarians, allowing patents on plants created using smart and conventional breeding techniques would upset the balance between public and private interests; the monopolies created by these patents, they argued, would hurt society in unacceptable ways. While the EPO ultimately allowed patents on these plant products but not on the processes,[139] this letter was another signal that the European Parliament had decided that patents shaped the socioeconomic implications of technology, and that Europe's governing institutions need to step in to address this.

The German parliament did step in. NPOS lobbied legislators in Germany, the Netherlands, France, and Switzerland to ban patents on conventional breeding methods and on the products derived by these methods.[140] Initially the German Bundestag issued a nonbinding resolution on the matter, but in June 2013, it revised its patent laws to prohibit patents on processes and products related to conventional breeding methods.[141] These laws covered decisions made by the German Patent Office, but did not cover EPO decisions. Some of the other legislatures studied the issue but have thus far taken no legislative action.[142]

This categorical exception in German patent law echoed those in the European Patent Convention and the BPD, as well as those covering food and pharmaceuticals in the nineteenth century. This approach suggested that any patent related to conventional breeding methods was so likely to be problematic that it warranted a proactive ban, which was even more forceful than the post hoc compulsory licensing provisions adopted in the case of human gene patents.

As the debates over patents on human genes and plants proceeded and were resolved, it became clear that the EPO and other European governing institutions understood patents as having important distributive effects. They saw the patent system as playing a role in shaping the socioeconomic impacts of technology. Embedded in this understanding was the idea that patents, and patent-based monopolies, should not be distinguished from one another. And as the EPO refused to address these issues in their examination and appeals processes, and limited the interpretation of the *ordre public* clause specifically, other parts of the pan-European patent system stepped in to take on these responsibilities.

CONCLUSION

This chapter has underscored the importance of an institutional and political approach to understanding the patent system. A purely legal analysis of the human gene and plant patent cases likely would have led us to conclude that the new wave of concerns regarding the distributional implications of patents has had a negligible impact in both the United States and Europe. We might also form an incorrect view of the *AMP v. Myriad* case as demonstrating a sea change in how the US patent system deals with citizen concerns or understands the public interest. But here I have demonstrated that legal outcomes are only partial indicators of complex institutional and political environments, which must be analyzed in broader context in order to make better sense of how the US and European patent systems function and to understand their roles and responsibilities, especially as they experience growing scrutiny.

What we see is that the United States has built a governance approach that supports its understanding of patents in techno-legal terms and, at a deeper level, supports a market-making ideology. In both the human gene and plant-patent cases, plaintiffs were essentially forced to use the courts to address their grievances. And the rules of the court, from case precedent to legal-standing requirements, made it impossible to see patents as having distributional implications. Even when the PTO was forced to contend directly with such concerns, it was unable to do so. Overall, we saw that the techno-legal approach to patents in the United States led to rhetoric, processes, and programs that define patents in narrow terms and the patent sytem as focused on economic growth by promoting inventors' interests.

In Europe, where the EPO had struggled for years to balance its commitment to make systematic decisions based on science and the law with its long-standing view of patents as moral and policy objects, we saw patent-system institutions respond to distributional concerns outside the scope of individual patent decisions. This approach fit with a market-shaping ideology that placed much greater faith in government, particularly in the context of innovation, than did its US counterpart. Thus, while the EPO refused to consider socio-economic implications within its definition of *ordre public*, it simultaneously tried to change its bureaucratic culture to respond to these concerns. Other European patent-system institutions began to take on more responsibility with regard to socioeconomic effects. The European Parliament increased its scrutiny—although still only informally—of the EPO. National parliaments also began to take an active role, passing laws that moved Europe further away from the harmonized, techno-legal approach to patents embodied in the EPC and dozens of international treaties. In sum, the European patent system was far more sympathetic to distributional concerns than EPO decisions on individual patents would suggest. By the 2010s, despite the many legal and cultural ties that bound them, the US and European patent systems operated quite differently. Catalyzed by the life-form patent controversies, they now experienced different political pressures, they had institutionalized rather different approaches to patents and their governance, and they had oriented themselves differently within the world of innovation.

CONCLUSION

Patent systems seem to operate far from the world of politics. We know that they influence innovation not only by providing the incentive of exclusive commercialization rights but also by creating legal hurdles that shape the direction of scientific inquiry. We see that they shape markets by influencing the goods that are available and the competition among producers. Despite these social effects, the systems that grant patents, and patents themselves, are usually treated as legal, technical, and relevant only to a specialized few. As patents on indigenous knowledge, essential medicines, and life forms have become controversial, these battles are usually characterized as bringing public-interest or political considerations into a domain otherwise ordered strictly by science and the law. This book shows otherwise.

PATENTS AS SOCIAL AND POLITICAL ACHIEVEMENTS

Throughout, I have demonstrated that patent systems are deeply political. Political culture, ideology, and history penetrate them, including their supposedly technical bureaucracies, just as they influence other nontechnical policy domains. This political context influences the social meaning of patents in substantial ways. In the United States, a market-making ideology combined with the emergence of robust expertise barriers surrounding the patent system have reinforced the definition of patents as limited techno-legal objects as well as innovation and market drivers. In contrast, a market-shaping ideology in Europe, reinforced by repeated distinctions between public and private interests throughout the history of its patent systems, has led to an understanding of patents as not only innovation and market drivers but also moral and socioeconomic objects. Consider the case of human gene patents. In both the United States and Europe, patents on the BRCA genes stimulated concerns about the implications for research and access to health care, and in both places these patents were ultimately revoked. But in the United States, traditional patent-system participants rejected the idea that patents had these distributive implications at all. Such impacts, they argued, were the responsibility of patent holders and not connected to patents themselves. Instead, the Supreme Court made the decision on the basis of legal precedent: the "product of nature" doctrine. In Europe, decision makers agreed with opponents that patents themselves could hurt research and access to health care, but disagreed about where and how this issue should be addressed. The EPO refused to incorporate this consideration in its examination practices, revoking the BRCA patents on other grounds. But the EPO has developed bureaucratic initiatives to sensitize its examiners to these kinds of public-interest concerns, and national governments took steps to address distributive concerns through compulsory licensing provisions in their laws.

We have also seen how rhetorical and institutional attempts by patent systems to distinguish the technical from the political are, themselves, political. These politics are simply features of these systems, not demonstrations of bias. This was clear in how the United States and Europe handled the controversies over animal patents. In the United States, traditional patent-system participants, including decision makers, characterized the patent system as a technical and administrative law domain. Concerns about the commodification of life, or the morality of producing genetically engineered animals, were to be addressed in research and regulatory domains, including through NIH and FDA rules. But what was treated as technical in the United States was defined as moral in Europe, where the European Union's BPD required patent offices to conduct an ethical assessment that weighed the suffering of the genetically engineered animals against the benefits to humankind. European decision makers saw patents, particularly in the area of biotechnology, as important moral certifications, sanctioning particular social practices as well as particular avenues of research and technology.

At the same time, these long-standing political orientations shape how we understand appropriate patent governance, and the appropriate governance of science and technology overall. Throughout the life-form patent controversies, the US patent system reiterated its understanding of the inventors' interest as synonymous with the public interest. With this understanding, it identified relevant knowledge and expertise in terms of science, the law, and the experiences of inventors and patent lawyers. Civil society perspectives were generally considered irrelevant—and even polluting—as their interests were already captured in the interests of inventors. As a result, its institutions had a limited set of responsibilities: they had to make predictable and procedurally objective decisions on the basis of science and the law, in order to facilitate the efficient introduction of new technologies into the marketplace. In contrast, when confronted with civil society challengers, European patent-system institutions returned to an earlier distinction between the interests of the public and those of inventors. So, they allowed average citizens and organized civil society groups to participate in the process, giving them opportunities to air their concerns. And they took these concerns seriously, forcing the EPO to wade—uncomfortably, to be sure—into new forms of evidence and expertise, including moral philosophy. European patent-system institutions then reimagined themselves as mediators, considering and adjudicating among multiple forms of knowledge, expertise, and participants. Indeed, while they emphasized procedural objectivity and respected scientific evidence and legal reasoning, they also began to value public responsiveness. But they began to see themselves as part of a larger apparatus for governing science and technology too, rather than as a distinct techno-legal domain far removed from questions about the broad implications of patents and innovation.

As a result, today we see that rather different constituencies, focused on somewhat different concerns, have arisen around the US and European patent systems. For example, while the destruction of human embryos for the purposes of hESC research provoked great controversy in the United States, these concerns barely registered in the patent system. Indeed, by then concerns about the ethics of commodification through patents had largely disappeared in a domain resistant to understanding patents as more than techno-legal or economic objects. Instead, the debate over hESC patents focused on whether they might stifle innovation, and it relied on data from scientists, economists, patent lawyers, and, to some extent, other social scientists. In Europe, by contrast, the European Parliament, EPO, and the ECJ saw patents and the patent system as shaping the moral status of both the human embryo and hESCs. This began when the European Parliament outlawed patents on inventions that involved the commercialization or exploitation of human embryos. With this move, it expected patent-system institutions to take the responsibility of shaping the marketplace to reflect the morality of the European community. Eventually, the EPO, and then the ECJ, interpreted the legal language to prohibit hESC patents, and as they did so, they also made room for ethical reasoning and expertise in the patent system.

These findings add a new dimension to our understanding of the long arm of politics, to be sure. They also provide new ways of thinking about science and technology policymaking, and particularly patent-system reform, in the context of growing citizen scrutiny. First, by revealing the moral and political orders embedded in our patent systems, they give us the opportunity to think critically about the values and assumptions that lie beneath them. Our understandings of what the scope of the patent system *is*, what questions are *relevant*, who should participate and how they should do so, and what role patent systems should play in the governance of emerging science and technology are not set in stone but are shaped by politics and society. The political cultures, ideologies, and histories that drive our patent systems are certainly fundamental and important, and therefore not infinitely malleable. But showing these underpinnings, particularly in comparative perspective, offers us more levers for reform. At the very least, it should give us pause when perspectives are taken off the table because they are deemed "irrelevant" in a limited "technical" domain.

Second, by forcing us to consider what lies underneath our distinctions between the technical and the political, and between the governable and the ungovernable, this book should lead us to rethink our approaches to relevant knowledge and expertise in science and technology policymaking. Knowledge for policy is not self-evident, and both policymakers and citizens must think critically about how their understanding of a policy problem intersects with their definitions of relevant knowledge, expertise, and participants.

Third, this analysis should remind us that despite the power of law, particular political and institutional environments give it meaning and importance. None of the conclusions drawn above are evident from our traditional method for analyzing patents, which focuses on legal language, reasoning, and precedent. In particular, focusing on the *ordre public* clause to explain the divergent paths of US and European patent law in the area of biotechnology misses much larger and more deep-seated differences in the politics and moral understandings of the two patent systems. If we take these differences seriously, we must rethink our efforts to harmonize patent systems and develop international trade agreements. Governments across the world, perhaps most notably in the United States and Europe, are investing huge amounts of time and energy into developing international legal agreements that will structure the global marketplace. This book suggests that even if national negotiators come to agreement and manage to harmonize intellectual property laws further, true harmonization is likely to be impossible. Jurisdictions will still make sense of these agreements in the context of their own distinct political cultures, ideologies, and histories, which will have enormous implications for their innovation, for their markets, and for their moral and political orders.

Finally, these conclusions should encourage us to reconsider our governance of the moral and socioeconomic implications of science and technology. The battles over life-form patents, as with many of the battles over science and technology policy, make plain a regulatory gap that continues to foster citizen unrest.

REFORMING THE PATENT SYSTEM

So how should we incorporate these findings into ongoing discussions of patent-system reform? As we have seen, innovators have begun to complain that patents can stifle research and technology development. Citizens increasingly see patents not only as modulating their access to medical care and technology, but also as shaping their livelihoods, and the general public has become increasingly skeptical of the idea that patents serve the public interest. In response, reform advocates have suggested alternative innovation systems, including public-private partnerships; alternative intellectual property regimes, including prizes; and new incentives to license inventions more widely. But these interventions tend to be market-based, and they leave aside the governance of current patent systems. This not only limits our levers for reform but also constrains the way we think about reform itself.

This book forces us to step back and take patent-system politics into account in these discussions. It teaches us that the framing of patent-reform debates, the patent-reform proposals themselves, and our perspectives on the viability of possible interventions are the results of assumptions and values that are embedded in our patent systems but usually obscured. It also demonstrates that unless reforms include attention to the rules, practices, and institutions of patent systems, they will have limited and short-term impact. In order to institute reforms that maximize the benefits of patents and innovation, we must think critically about the logics that underlie patent systems, understand how these logics are integrated into patent-system operations (including examination processes), and then consider their malleability. To assist us in this process, we must ask ourselves: What do we want out of our innovation systems? What role can and should patent systems play in our innovation systems? How might different kinds of participants and discussions in our patent systems produce the kinds of innovation and innovation systems that we want? Only then can we begin to envision fundamental reforms that could magnify the benefits of innovation and patents for all of us.

This book also illuminates how the growing distrust facing patent systems echoes citizen concerns regarding science and technology policymaking domains more generally. Citizens are no longer content to accept passively that these institutions operate in their interest because their decisions are systematic and based on science and the law. Instead, they question whether these decision makers reflect their values and priorities. They no longer accept that rewarding innovators—through research funding or through patents—will necessarily benefit the public. Increasingly, citizens expect government to be responsive to their concerns. This demand can be particularly difficult for governments to manage, particularly because these concerns usually do not fit neatly into existing decision making approaches. In addition, governments may characterize this push for responsiveness as being in opposition to its goals of efficiency and predictability. In this analysis, we have seen two different approaches to this challenge. Despite almost four decades of pressure, the US patent system has responded by reasserting its expertise barriers, including its procedural objectivity. It has emphasized its systematic and predictable decision making, and it has characterized the concerns of civil society groups as irrelevant, misguided, and sometimes even biased. It seems to operate under the presumption that by reasserting its approach, which promotes efficiency and a busy patent system, the concerns will eventually dissipate. The pan-European system, by contrast, has slowly become more responsive to public concerns, but it has also tried to systematize this responsiveness. Largely in response to the life-form patent controversies, it is now more open to civil society critique and a range of expertise, and it has developed a broad understanding of social implications. Its has formalized this attention through interventions such as the SeCa system and systematic attention to patent critics. As a result, it issues fewer and narrower patents and is less efficient, especially in more controversial areas of technology. Innovators, as a result, may be underserved, but there is no evidence that there is less innovation as a result. And while the EPO continues to experience civil society opposition as well as scrutiny from the European Parliament and other European institutions, it seems somewhat more comfortable performing this balancing act.

In both the United States and Europe, not to mention the rest of the world, we continue to see considerable controversy over patents. This suggests we haven't yet struck the right balance in producing decision making that is simultaneously systematic, efficient, and responsive. As the United States and Europe develop solutions, they can certainly look to other jurisdictions for ideas, but as they do so, they must take into account the distinct political cultures, ideologies, and histories that shape patent systems. The United States, for example, cannot simply implement the European approach; rather, it must assess potential policies and decision-making approaches with its cultural and ideological commitments in mind. Patent-system critics in both the United States and Europe benefited from watchdogs who possessed hybrid expertise. But transforming the US patent bureaucracy to make it more accepting of nontechnical expertise is difficult to imagine in a jurisdiction that values procedural objectivity so completely. However, in a jurisdiction predisposed to adversarial debate and to incorporating formal public comments into bureaucratic decision making, perhaps there should be more explicit and formal attention to the concerns of the public beyond those in the innovation community. Or, given US support for public-interest litigation, it might require an explicit expansion of legal-standing requirements or institutional support of public-interest cases. In Europe, reform may require standing ethics advisors or committees within the EPO to complement its chief economist. It may also involve more systematic rethinking of how patent examiners are trained. This might include development of formal decision-making guidance to supplement its otherwise legally focused patent-examination manual, to ensure that examiners can make judgments based on nonquantitative evidence, such as ethical reasoning or socioeconomic data. And if the EPO continues to see organizations like Kein Patent Auf Leben as institutional watchdogs, then these watchdogs may need more formal support.

RETHINKING INNOVATION GOVERNANCE

Even if we decide that we are happy with current patent systems and believe that they are not the place to address growing moral and distributional concerns over science and technology, our work is not over. This book has suggested that citizens are deeply concerned about who owns technology and about the power that this ownership confers. This includes worries that our current governance systems actually inhibit, rather than expand, access to technology. Furthermore, the citizen critiques I have highlighted throughout this book challenge basic assumptions about how innovation and innovation systems work. They call into question, for example, the idea that everyone will eventually benefit from technological development. For the most part, governments have ignored these concerns. To the extent that they have acknowledged citizen frustrations related to science and technology, they have focused on issues of health and environmental risk. This is understandable: on their face, questions of risk and safety seem easily answered with quantitative data and objective measures of evaluation.

Patent systems have become a target because of their early and formative involvement in shaping innovation trajectories and because of their roles in shaping commodities and markets. But if we are convinced that the patent system is the wrong place to address these concerns, then policymakers, scholars, and citizens must figure out where we *do* want to address them, and we must develop the structures to address them systematically and proactively. Otherwise, we risk continued and perhaps increased citizen unrest over science and technology, which poses challenges for both innovation and society.

We must do more than rely on bioethics and other advisory committees that serve largely to inform public debate. We must focus on the challenge of incorporating these concerns into our policy decisions. We do have some models to guide us. The United Kingdom, for example, has incorporated explicit attention to ethical concerns into bureaucratic decision making related to reproductive technology and embryo research.[1] It does this with extensive efforts to solicit public comment, as well as through the use of several ethics advisory committees whose recommendations affect policymaking directly. In addition, scholars of science and technology policy have been experimenting with different models of public deliberation to inform decision making.[2] Still, we have developed few policy frameworks so far to address systematically the distributive concerns that have arisen in the patent controversies.

As societies and citizens consider how to reform their science and technology policies to consider these issues, they must think critically about the kinds of knowledge and expertise we include and how we decide to value it. The findings in this book encourage us to scrutinize our assumptions about best practices for evidence-based policymaking. We often hear that good policymaking, particularly in highly technical policy domains, must rely on "sound science." Put differently, if only policymakers would rely on more evidence, they would make better decisions. But what these calls omit is that our understanding of what constitutes knowledge and evidence, not to mention our methods of evaluation, are shaped by values and politics. We saw, for example, how the governing institutions in the European patent system treated ethical reasoning, knowledge related to international development, and social-science expertise as relevant, in contrast to their US counterparts. In the United States, patent-system institutions and organized interests used a variety of strategies to maintain a narrow understanding of relevant knowledge and expertise in the domain. When it comes to science and technology policy and politics more broadly, then, it is not just a matter of bringing in more evidence or clarifying which questions are scientific and which ones are about values. Rather, we must also become aware that our understanding of what constitutes "scientific" or "ethical" in a particular policy domain is the result of particular norms, formal and informal rules, history, and dominant ideologies.

In addition to helping us develop more publicly legitimate approaches to considering the moral and socioeconomic implications of science and technology, this awareness should help governments and citizens alike better interpret and navigate science and technology policy controversies. Consider, for example, the emerging discussion about the regulation of CRISPR/Cas9.[3] This new technique is seen as revolutionary because it allows scientists to cut out and replace pieces of the genome more efficiently and possibly more accurately than by using previous genetic modification techniques. It has already been used to alter the genomes of plants and animals, and scientists have begun to conduct experiments on human embryos. This manipulation, particularly because it will be passed down to future generations, has generated considerable media attention and stimulated controversy. In response, respected organizations, including the National Academies of Science in the United States and the Nuffield Council of Bioethics in the United Kingdom, have convened expert committees to make policy recommendations to their respective governments and scientific communities. They are, in other words, assembling truth to inform power. But what constitutes "good" knowledge to inform CRISPR research policy? How can we ensure that average citizens benefit from the new technology while bearing a minimal burden of risk? Policymakers must resist their impulses to turn to scientists and engineers as the primary sources of expertise. Social scientists and humanists offer a nuanced understanding of the moral and social implications of biotechnology that is pivotal for governments navigating these issues. They also have important insights to offer on the governance of innovation from historical and comparative perspectives, which can help us illuminate the otherwise taken-for-granted assumptions and values embedded in our science and technology policies. Meanwhile, citizens can offer unique expertise beyond simply providing the pulse of public opinion. They can teach policymakers in these domains about their experiences with and expectations of innovation, helping us to maximize public benefits. In sum, in order to achieve better science and technology policies, we must learn how to embrace and engage their political dimensions. Only then can science and technology achieve its social potential.

ACKNOWLEDGMENTS

During a break in a 2004 EPO opposition hearing on the BRCA1 gene patents, one of the opponents—a European scientist—turned to me and said, "You can force us to go to war in Iraq, but you cannot force us to accept your gene patents." This statement, from a relative stranger, took me aback (I had interviewed him once, a few months before). It stayed with me, sparking questions that would nag at me for years. What did the Iraq war, and geopolitics more generally, have to do with patents? Why was this scientist thinking about these issues during this seemingly technical review process? And why were these politics so present to this scientist that he wished to express them to me at that moment? These questions ultimately sent me on a long journey into the patent system.

Along the way, I was supported by dozens of people and institutions to whom I owe enormous thanks. First and foremost, I am deeply indebted to interview subjects in both the United States and Europe, who were so generous with their time and their reflections. Through their words, they brought seemingly esoteric patent systems to life and taught me what was at stake for them—and for us—in the life-form patent controversies. They also provided color and richness to the histories that I was reconstructing through other means. Indeed, archivists—both formal and informal—were immensely helpful, especially Päivi Vainiomäki at the European Parliament's archives and Joyce Bichler at the Animal Legal Defense Fund.

I could not have done the time-consuming and labor-intensive research that informed this book without additional research support. I began this project with seed grants from the Center for Ethics and Public Life and the Center for International Business Education at the University of Michigan. Fellowships from the American Council of Learned Societies and the Woodrow Wilson International Center for Scholars allowed me to take research leave from my faculty position so that I could go into the field, and live between Washington, DC, and Munich, Germany for almost three years. While in Germany, I received financial assistance from the Max Planck Institute for Intellectual Property, Competition, and Tax Law. The Science, Technology, and Society Program at the National Science Foundation deserves special thanks for providing, through a Scholar's Award (SES-0724664), the bulk of the funding for this project. It ensured that I could leave no stone unturned. A sabbatical year at the American Bar Foundation and funding from University of Michigan's Risk Science Center gave me the space and time to analyze my findings and begin to draft the book. I particularly want to recognize the vibrant scholarly communities created by both the Wilson Center and the American Bar Foundation, who fostered my intellectual curiosities and helped me maintain joy in the work. I am also thankful to the Ford School of Public Policy. Hillary Pryor and Sharon Disney provided essential administrative help every step of the way. Dean Susan Collins, Associate Dean Paula Lantz, and the Ford School community were enthusiastic supporters. I simply could not have done this work without my veritable army of research assistants: Monamie Bhadra, Alex Farivar, Claire Giammaria, Molly Maguire, Erica Morrell, Anita Ravishankar, Katie Reeves, Alexis Walker, and Caroline Walsh. I owe particular thanks to Monamie, Alexis, and Caroline for their tenacity, keen critical eyes, and willingness to pitch in whenever and however I asked, not to mention their friendship.

Interpretive projects such as this one are constantly being rethought and reworked until the meaning of the data comes into sharp focus. In the process of analyzing and presenting your findings, you discover new angles on and understandings of the data. Then you rewrite, reanalyze, rediscover, and repeat. While much of this work is solitary, it is influenced deeply by the communities within which it is produced. I benefited greatly from giving talks at Cornell University, Harvard University, Northwestern University, Arizona State University, New York University, the University of California at San Diego, the University of Michigan, Virginia Tech, the University of Vienna, the University of Augsburg, the University of Waterloo, the University of Oslo, Tel Aviv University, the Indian Institute of Management at Bangalore, Gordon Research Conferences on Science and Technology Policy, and meetings of the Society for the Social Studies of Science. Discussions with John Carson, Benjamin Cohen, Jason Delborne, Daniel Diermeier, Ulrike Felt, Tim Forsyth, David Guston, Ben Hurlbut, Pierre-Benoit Joly, Anna Kirkland, Brice Laurent, Clark Miller, Jessica Palmer, Roopali Phadke, Jody Roberts, Shep Ryen, Dan Sarewitz, Perrin Selcer, and David Winickoff greatly enriched the project as well. I am so happy to call them not just colleagues but also friends. Di Bowman, Rick Hall, and Jim House read and commented on my work in its early stages and provided pivotal insights. The Science and Democracy Network community offered valuable perspectives on numerous occasions as it saw the project evolve. Finally, I feel very lucky to have had the continued guidance and encouragement of Sheila Jasanoff and Stephen Hilgartner, who provided intellectual clarity in the moments that I needed it the most.

Kelly Moore, Alexandra Murphy, Alex Ralph, and Joy Rohde were wonderfully supportive friends, but they also volunteered to perform the extraordinary task of reading and commenting carefully on parts of or the full manuscript (in some cases, more than once). They offered invaluable constructive criticism that kept me motivated, especially during those dark days of revision, and I hope that I have managed to do it justice. I am very grateful to my editor, Karen Merikangas Darling, for her investment in the book, her tireless work on my behalf, and her patient answers to all of my questions. Thanks also to Evan White, Nicholas Murray, Julie Shawvan, and Mary Corrado for their help with the production process. And to the anonymous reviewers of this manuscript, thank you very much for your detailed and incisive comments and for helping me to understand what this project could be. Even with all of this assistance, however, any errors or shortcomings in this manuscript are solely my own.

With their encouragement and love, my friends and family have sustained me over the many years of researching and writing this book. I value deeply my lifelong friendships with Carla Bittel, Ruth Farrell, Mona Jhawar, Lucy Stanfield, and Jamey Wetmore, and I look forward to many more years of laughter, indulgence, and adventure. Evelyn Alsultany, Deirdre de la Cruz, Mayte Green-Mercado, and Mara-Cecilia Ostfeld, who have become my Ann Arbor sisters, lent me their ears, their hearts, and their empathy. Josh Johnson, Tom Woods, Julie Epton, and the gym crew kept me well balanced, with my eyes on the prize and a smile on my face. Finally, I am incredibly fortunate to have such an engaged and caring family. My parents, Padmaja and Partha Sarathy, and my brother Suhas, have mastered the delicate dance of fostering my ambitions while making sure I keep it all in perspective. Thank you for your unwavering love and belief in me.

APPENDIX 1
MAJOR EVENTS RELATED TO THE US AND
EUROPEAN LIFE-FORM PATENT CONTROVERSIES

1623   English Statute of Monopolies established.

1790   United States passes its first patent law.

1817   US Supreme Court articulates moral utility doctrine in *Lowell v. Lewis* case.

1883   Paris Convention signed.

1963   Strasbourg Convention signed.

1972   Ananda Chakrabarty applies for a US patent on a genetically engineered microorganism.

1973   European Patent Convention signed.

1980   US Supreme Court decides *Diamond v. Chakrabarty* case.
        US Congress passes Bayh-Dole Act.

1988   European Commission introduces Biotech Patent Directive (BPD).
        US PTO Commissioner Quigg issues memo regarding animal patents.

1991   Craig Venter applies for US patents covering cDNA.
        Animal Legal Defense Fund files lawsuit against Quigg memo in a US district court.

1992   Europe approves initial version of Oncomouse patent.

1995   European Parliament rejects BPD.

1998   European Parliament passes revised BPD.
        Jeremy Rifkin and Stuart Newman file US patent application on human-animal chimera.

1999   German Patent Office awards Oliver Brüstle patent on neural stem cells.

2000   EPO issues "Edinburgh" patent.

2001   EPO issues patents on the BRCA genes to Myriad Genetics.
        US PTO issues new "utility" guidelines in response to the cDNA patent controversy.
        US President George W. Bush issues executive order restricting federal funding for hESC research.

2002   EPO Opposition Division rejects Edinburgh patent.

2003   Europe approves revised version of Oncomouse patent.

2004   California stem cell initiative passed by voters.

Christoph then challenges German Brüstle patent.

France approves amended law authorizing compulsory licensing for gene patents.

2006   CIRM Task Force introduces novel intellectual property policy.

Public Patent Foundation asks the US PTO to reexamine WARF/ Thomson patents.

2007   EPO introduces Raising the Bar initiative.

EPO releases *Scenarios* report.

2009   ACLU and PubPat file *AMP et al. v. Myriad and USPTO* case in US district court.

EPO Enlarged Board of Appeal rejects WARF and Edinburgh patents.

2010   US SACGHS issues report, *Gene Patents and Licensing Practices and Their Impact on Patient Access to Genetic Tests*.

2011   European Court of Justice resolves *Brüstle v. Greenpeace* case.

United States passes America Invents Act.

2012   US District Court judge rejects OSGATA case.

2013   US Supreme Court decides *AMP et al. v Myriad* case.

German Bundestag prohibits patents on processes and products related to conventional breeding methods.

2015   US PTO dismantles SAWS program.

APPENDIX 2
A METHODOLOGICAL NOTE

This book is a comparative, interpretive analysis of the politics of the US and European patent systems. Focused on the debates over life-form patents, it is based on multiple sources of qualitative data: archival and current documents, interviews, and participant observation. It focuses on the life-form patent controversies both because they have lasted for more than thirty years and because they have touched upon many of the concerns that have been raised regarding patents related to other technologies as well. As a result, the book gives us insights into the broader politics of intellectual property and of science and technology policy. In this appendix I provide an overview of the research I conducted and my analytic approach. Some of the sources I used have not been analyzed by humanists or social scientists previously, and may prove useful for future analysts of patent systems.

ANALYZING THE LIFE-FORM PATENT CONTROVERSIES

Research for this book began in late 2006. After attending an EPO opposition hearing against one of Myriad Genetics' BRCA patents in 2004, as part of the fieldwork for my previous book, I became increasingly curious about the politics of life-form patents. The participation of civil society groups at the EPO opposition hearing and their opportunity to raise moral, social, and distributional concerns—in what seemed to be a highly technical and legal forum—astonished me. It provoked me to investigate these kinds of civil society–led efforts further and to try to understand whether the life-form patent controversies were unfolding similarly in the United States and Europe.

Document Analysis
My first step was to understand the scope and history of the life-form patent debates in the two places. I pursued this on several fronts. I looked at news sources, using the LexisNexis Academic database, and databases for individual newspapers (e.g., *Le Monde*, *Süddeutsche Zeitung*, the *Financial Times*, the *New York Times*). I searched for relevant articles from the 1970s to the 2000s, using different combinations of search terms such as *patent, gene, life form, biotechnology*, and *controversy*. (As I became more familiar with the major events and issues, I varied the search terms a little, adding terms related to specific cases,

including *stem cell*, "Myriad," and "Oncomouse"). I read these articles (which eventually numbered in the thousands) in order to identify the major events, decision-making institutions, and players and issues in these controversies. In time, I also used these news articles to understand the types of rhetoric that participants used throughout the debates and to consider how they compared across jurisdictions. Once I had identified the governmental and nongovernmental participants in these controversies, I visited their Web sites and read (and often downloaded) the information they presented. This included court, civil society, and bureaucratic (including patent-office) documents. This helped me develop a more detailed history and understanding of the controversies that had taken place. It also provided me with primary documents that I could use to analyze the strategies of the various players involved. (My interviews and fieldwork trips also often allowed me to gather similar information—including press releases, position papers, and reports—that I could analyze in a similar manner and use to cross-check. When I visited the PTO and the EPO, for example, I also went to their libraries and retrieved copies of their internal newsletters and annual reports.)

I collected additional information from databases supported by the PTO and the EPO (Public PAIR in the United States and Register Plus/Espacenet at the EPO), about the life-form patents being challenged by civil society groups. Using information I had gathered from news and other online sources, I accessed database entries about the controversial patents themselves. These entries included timelines regarding the fate of the patent, lists of inventors and opponents (as relevant), and bureaucratic information that I could use to search for additional information. I then used these databases to search for other patents that had provoked official civil society challenges at the PTO or the EPO but had not garnered much public or media attention (once I identified frequent EPO opponents, for example, I used the EPO's database to search for other patents that those groups opposed.) Eventually, I assembled my own database that recorded information related to 142 life-form-related patents that had generated some kind of controversy in the United States, Europe, or both. I included patents that had been challenged at the patent offices, in the courts, in legislatures, via petitions, and on the streets. This investigation, coupled with ongoing media and online data gathering, helped me develop an understanding of the scope of the life-form patent battles, the players involved, and the fate of each challenge.

Most importantly, for those patents challenged directly at the PTO and EPO, these databases (and visits to the patent offices) gave me access to every "prosecution" related to a patent. These included all of the communications between the patent applicants, patent-office personnel (receiving section, examiners, opposition and appeals boards), and third parties who submitted documents relevant to the legal dimensions of the case (both official opponents and others). Interested parties can download parts of the prosecution or the full file, which provide public access to the initial patent application, the response from the examiner, communications between the applicant and examiner over appropriate wording of the patent, official oppositions and their supplementary documentation, third-party submissions, and the minutes and decisions of appeals board hearings. Some of the older patent prosecutions were only available in hard copy, so I retrieved them from the patent offices directly. Full documentation can run into the thousands of pages (the EPO's Oncomouse case is more than seven thousand pages long, for example). I downloaded and read the full files for the most controversial cases (about forty in total, including those discussed in the book). They are extraordinarily rich sources of data for analysts of the patent system. They allowed me to establish a more detailed understanding of the history of each case and helped me understand how each system worked at a much deeper level. They also gave me an opportunity to analyze the rhetoric invoked by all parties and the structure of the decision making processes. Overall, these prosecutions allowed me to reconstruct the social and political dynamics of these bureaucratic spaces.

I also accessed court documents where appropriate. This was particularly important for the United States, where most disputes occurred in courtrooms, but it was useful for Europe as well, as conflicts moved into national courts and the European Court of Justice. So I retrieved and analyzed the documents related to the cases discussed throughout this book, including submissions from the petitioners and defendants, amicus briefs, judicial opinions, statements of support, and expert declarations. I gathered data on other related court cases as appropriate.

Interviews

Seeking to interview as many people as I could who were involved in the life-form patent controversies in the United States and Europe, I identified my first group of interviewees through the initial document analysis. Many of these prospective subjects graciously agreed to be interviewed and suggested additional interview subjects. (As I continued my research, I also identified additional interviewees through continued document analysis and ethnographic observation.) Overall, I knew that I needed to interview the major (nongovernmental and governmental) players in the controversies. This included a large number of personnel at the US and European patent offices who had contended with life-form patents as examiners, supervisory examiners, public relations officials, legal counsel, and high-level officials. I managed to do this during extended fieldwork stays in Washington, DC, and Munich, where the US and European patent offices are respectively based. These stays were supplemented with trips across the United States and Europe to conduct interviews with other government officials and players in the controversies.

Overall, I interviewed more than 100 individuals in the United States and Europe, including 45 current and former personnel at the PTO and EPO; 28 individuals affiliated with the patent challenges in the United States and Europe, including but not limited to representatives of Greenpeace, KPAL, NPOS, PubPat, the ACLU, and ALDF; 9 individuals representing groups who generally supported expanded life-form patentability, including but not limited to the US, UK, and European biotechnology associations; 17 US and European patent lawyers who had represented either patent applicants (or owners) or challengers; 11 US and European scientists who had applied for one of the contested patents; and 13 US and European government officials outside the patent system, including but not limited to bureaucrats inside the US Department of Commerce and European Commission's Directorate related to industry, parliamentarians of the European Union and Germany, and congressional staff members in the United States.

I conducted in-person, semistructured, and open-ended interviews with each person. For each interview, I sought to cover a set of themes that I had predetermined, although the flow of each interview differed in order to maximize the subject's ease. This technique helped me to be sure that I covered all issues of particular interest to me while also providing flexibility to tailor the interview according to the responses of the interviewee. It also allowed me to conduct the interview in the form of a conversation; this was particularly important, I thought, because it might elicit more genuine responses and diminish the awkwardness of a formal interview. I was generally interested in learning about the individuals and how they chose their profession, what they could tell me about how their group or institution operated, their involvement in the politics of life-form patents, their reflections on what these politics looked like and how they had evolved, why they thought life-form patents had become so controversial, how they envisioned change, and what the future held for the patent system. Depending on the cases that they were involved in, I also pressed them for more specific details and reflections. Finally, I asked each interviewee for suggestions about others to interview, other documents that I should read, and events that I should attend. These interviews generally lasted about an hour, but in many cases they lasted longer, or I scheduled one or more follow-up interviews.

Each of these interviewees signed an informed consent form that described the purpose of the study and how I planned to use the interview data. In the consent form, I stated that I would do my best to maintain the interviewee's anonymity. While it may be difficult to believe that this type of social science research has any potential risks, I recognized that given both the political and economic dimensions of the life-form patent debates, subjects might be less likely to speak freely if they thought their statements could be easily traced back to them. In the consent form I also noted that if I found it impossible to use an interviewee's statements without making his or her identity obvious, I would seek additional permissions before publishing anything.

Once I had transcribed these interviews, I looked for common themes among them. I asked whether there were clear similarities or differences between US and European interviewees in terms of how they thought about life-form patents and their politics, and whether there were more specific similarities or differences among the different classes of individuals that I interviewed.

Participant Observation

I supplemented document analysis and interviews with participant observation of these players and conflicts in action. I attended, for example, two EPO opposition hearings, an EPO appeals board hearing on the WARF stem cell patent, and the US Supreme Court hearing on human gene patents. I also witnessed a few street protests along the way. I attended less controversial meetings and conferences, including a town hall held by the PTO, a workshop on civil society involvement at the EPO, and meetings of the Secretary's Advisory Committee on Genetics, Health, and Society, as well as the European Policy for Intellectual Property association. Attending all of these events provided me with a deeper understanding of both the politics of life-form patents in action and how these politics fit into the broader dynamics of the patent system. I could observe how players in these politics interacted with one another and understand how nonverbal elements (e.g., body language, the location, architecture, and even furniture arrangements) might shape these contentious politics.

Access Challenges in Comparative Perspective

In many respects, it was equally easy to conduct research in the United States and Europe. The databases that I used to gather data were either publicly available or straightforward to access through the University of Michigan or other organizations with which I was affiliated. I was able to attend legal proceedings and other events in both places, and, for the most part, people in both countries were equally willing to speak with me.

There was one major difference, in terms of access to the PTO and the EPO, which warrants reflection. I had no trouble interviewing EPO personnel and working in its archives. But my experience with the PTO was quite different. When I first began contacting individuals for interviews, using publicly available information to do so, I received no response. After a few weeks, I eventually received a response from someone in the PTO's public relations office. Someone had forwarded my request to her, and she said that she would be happy to set up interviews on my behalf if I would tell her the types of people I wanted to interview. I gave her an initial sense of whom I wanted to speak with, and over the next eighteen months, she set up meetings for me and also accompanied me to each interview. This was great. She provided me with substantial access (when I last saw her, she said that I'd had more access than any other previous researcher). But, she also directed with whom I spoke and, sometimes, for how long. And while she was silent during my interviews, and I was free to ask whatever I wanted, I was afraid that her presence might influence the interviewees' responses (although I saw no evidence of this).

I addressed these constraints by pushing the boundaries: I made requests for interviews with specific people, asked interview subjects for other contacts in her presence, and asked if interviewees might be able to speak for a longer period than the allotted time. These efforts were usually successful. I also doubled my efforts to interview former PTO employees, including high-level officials, beyond this public relations official's purview. This was also successful. I did not find any significant difference between those interviews conducted with the public relations official's presence and those conducted without, which was heartening.

But the PTO put up another roadblock too. As I described in chapter 3, on a couple of occasions I submitted Freedom of Information Act requests in order to get access to historical documents and the SAWS Program. These requests were rejected, and I did not resubmit them, but instead used other means to get most of the documents I needed. While I resolved the immediate access challenges, it is worth noting that the PTO's responses to my inquiries reflect some of the comparative points that I've made throughout this book. PTO officials are unaccustomed to and uncomfortable with this kind of scrutiny. They focus on direct services to inventors and their legal representatives, and on their role in stimulating innovation and markets. Other kinds of publics were simply not part of their purview, and they simply did not know what to make of me. Indeed, my interviews with both current and former personnel reflected this approach as well.

PUTTING THE LIFE-FORM PATENT CONTROVERSIES INTO HISTORICAL PERSPECTIVE

As I pieced together the history and details of the US and European life-form patent controversies, it became increasingly clear that I needed to develop a better understanding of the historical origins and evolution of the two patent systems. I wanted to explore exactly what the similarities and differences were, and where they came from. I discovered quickly that there were very few political histories of US or European patent systems (including national European patent systems). To analyze the evolution of the US system, then, I collected all congressional hearings and reports, and all appeals and Supreme Court cases (including amicus briefs), related to patents from the eighteenth century to the present. For its pan-European counterpart, I gathered documents from the archives of the Council of Europe, the European Economic Community, the European Free Trade Association, the European Union Parliament and Council, and the EPO archives. I also found primary and secondary sources that laid out the histories of how the national patent systems shaped the evolution and structure of the pan-European patent system.

I used these sources to identify the major events, themes, and participants in the histories of the two systems (and inevitably, with comparison, to consider the factors that were unique to each system and to the historical periods.) For example, I paid close attention to how the public interest was invoked and defined, and by whom, in the two jurisdictions. I noted, for example, that until the *Chakrabarty* case, inventors and patent lawyers submitted almost all of the amicus briefs to the US Supreme Court, with no participation from public-interest groups who were not financial stakeholders in the system.

AN INTERPRETIVE AND COMPARATIVE APPROACH TO A POLICY DOMAIN

As I collected this qualitative data, I "triangulated" it in order to develop an in-depth understanding of the US and European life-form patent controversies.[4] This approach allowed me to observe convergence and corroboration across sources and types of evidence, and thus develop a nuanced understanding of the history of the life-form patent controversies in the United States and Europe. But the process was also iterative. As I began to analyze the data from various sources, certain relationships began to appear important. These preliminary patterns in turn led me to focus on particular types of data and particular sources, with which I could validate or reject the importance of the relationships I had uncovered. For example, I would compare an interview subject's interpretation of events with the interpretations recorded in the media or recollected by other interview subjects. The discrepancies often provided insight into alternative explanations, which I then investigated using less subjective sources. Or the differing interpretations themselves became important for unpacking what happened and why. Of course, what appears important early in a project often proves unfruitful as one goes into greater empirical depth. New findings didn't fit the working narrative; I needed to develop a new narrative that did fit and then subject it to new scrutiny, often using different sources. The challenge of this research method is to bring conceptual order to complex processes, moving back and forth between empirical investigation and interpretive understanding. The purpose of this research is not to abstract away from particular cases in search of general explanations that hold across time and context, though this kind of work is good at identifying the mechanisms that may inform efforts at general explanation.

This inductive approach aims to understand and analyze the complex structures of meanings and values through which actors see the world. It helps to develop the categories that give a complex reality conceptual order, and it clarifies the questions that require scholarly attention. It also begins the process of explanation by identifying and refining hypotheses and concepts and for understanding the complexity of social systems and processes; it can provide detailed understandings of the variables and relationships that are important, and of the ways social mechanisms work.

The comparative lens is designed to enhance the analytic edge of my interpretive work by allowing differences to emerge among cases; these differences form the basis for systematic analysis of similarities, differences, and possible causal relations. The goal in this kind of comparative analysis is not to establish strict causality between independent and dependent variables. Rather, it focuses on uncovering which elements are most important in shaping a particular phenomenon, and how those elements shape the phenomenon. In my work, the comparative approach also often reveals alternative policy solutions for what may appear to be similar scientific, technological, and social dilemmas. However, as it does this, it also identifies the different standards and priorities, and occasionally the different understandings of best practices, that guide the development of policies in different contexts. Therefore, while it can offer alternatives for resolving a particular dilemma and motivate policy-makers to reflect on how and why they make particular choices, the comparative approach also shows that a policy's viability as a model depends on the standards and priorities of the specific context.

NOTES

INTRODUCTION

1 No Patents on Seeds, "European Patent Office a 'democracy-free zone'" (press release). Online at https://no-patents-on-seeds.org/en/information/news/european-patent-office -democracy-free-zone.

2 Karuna Jaggar, "From the ED: Historic Gene Patent Case and Rally at Supreme Court," *The Source* 118. Online at http://www.bcaction.org/2013/04/05/from-the-ed-historic-gene -patent-case-and-rally-at-supreme-court/.

3 Diamond v. Chakrabarty, 447 U.S. 303 (1980).

4 Graham Dutfield, *Intellectual Property Rights and the Life Science Industries: Past, Present, and Future* (Hackensack, NJ: World Scientific Publishing, 2009); Andrea Stazi, *Biotechnological Inventions and Patentability of Life: The US and European Experience* (Northampton, MA: Edward Elgar, 2015); Keith Aoki, *Seed Wars: Cases and Materials on Intellectual Property and Plant Genetic Resources* (Durham, NC: Carolina Academic Press, 2008).

5 Margo A. Bagley, "A Global Controversy: The Role of Morality in Biotechnology Patent Law," University of Virginia Law School Public Law and Legal Theory Working Paper Series, Paper 57 (2007). Online at http://law.bepress.com/cgi/viewcontent.cgi?article= 1097&context=uvalwps; Benjamin D. Enerson, "Protecting Society from Patently Offensive Inventions: The Risk of Reviving the Moral Utility Doctrine," *Cornell Law Review* 89 (2004): 685–720.

6 Adrian Johns, *Piracy: The Intellectual Property Wars from Gutenberg to Gates* (Chicago: University of Chicago Press, 2010); Robert P. Merges, Peter S. Menell, and Mark A. Lemley, *Intellectual Property in the New Technological Age*, 6th ed. (Aspen, 2012); Robert P. Merges and John Fitzgerald Duffy, *Patent Law and Policy: Cases and Materials*, 6th ed. (LexisNexis, 2013); Peter Baldwin, *The Copyright Wars: Three Centuries of Trans-Atlantic Battle* (Princeton, NJ: Princeton University Press, 2014).

7 Kathryn Steen, *The American Synthetic Organic Chemicals Industry: War and Politics, 1910–1930* (Chapel Hill: University of North Carolina Press, 2014); Steven W. Usselman, *Regulating Railroad Innovation: Business, Technology, and Politics in America, 1840–1920* (New York: Cambridge University Press, 2002).

8 Daniel Lee Kleinman, *Impure Cultures: University Biology and the World of Commerce* (Madison: University of Wisconsin Press, 2003); Philip Mirowski, *Science-Mart: Privatizing American Science* (Cambridge, MA: Harvard University Press, 2011).

9 Heidi L. Williams, "Intellectual Property Rights and Innovation: Evidence from the Human Genome," *Journal of Political Economy* 121, no. 1 (2013): 1–27.

10 Marianne de Laet, "Patents, Travel, Space: Ethnographic Encounters with Objects in Transit," *Environment and Planning D: Society and Space* 18 (2000): 149–68.

11 Bronwyn Parry, *Investigating the Commodification of Bio-Information* (New York: Columbia University Press, 2004); Marilyn Strathern, "Potential Property, Intellectual Rights and Property in Persons," *Social Anthropology* 4, no. 1 (1996): 17–32.

12 Cori Hayden, *When Nature Goes Public: The Making and Unmaking of Bio-Prospecting in Mexico* (Princeton, NJ: Princeton University Press, 2003).

13 Jonathan Kahn, *Race in a Bottle: The Story of BiDil and Racialized Medicine in a Post-Genomic Age* (New York: Columbia University Press, 2014); Stephen Hilgartner, "Novel Constitutions? New Regimes of Openness in Synthetic Biology," *BioSocieties* 7, no. 2 (2012): 188–207.

14 Daniel Kleinman, *Politics on the Endless Frontier: Postwar Research Policy in the United States* (Durham, NC: Duke University Press, 1995); Robert Bud, *Penicillin: Triumph and Tragedy* (New York: Oxford University Press, 2007); John H. Perkins, *Geopolitics and the Green Revolution: Wheat, Genes, and the Cold War* (New York: Oxford University Press, 1997).

15 Kelly Moore, *Disrupting Science: Social Movements, American Scientists, and the Politics of the Military, 1945–1975* (Princeton, NJ: Princeton University Press, 2013).

16 Rachel Carson, *Silent Spring* (New York: Houghton Mifflin, 1962); Mark Hamilton Lyle, *The Gentle Subversive: Rachel Carson, Silent Spring, and the Rise of the Environmental Movement* (New York: Oxford University Press, 2007).

17 Elizabeth Bomberg, *Green Parties and Politics in the European Union* (New York: Routledge, 1998); R. C. Mitchell, "From Conservation to Environmental Movement: The Development of the Modern Environmental Lobbies," in *Governance and Environmental Politics*, ed. M. J. Lacey (Washington, DC: Wilson Center Press, 1989); Riley E. Dunlap and Angela G. Mertig, "The Evolution of the US Environmental Movement from 1970 to 1990: An Overview," in *American Environmentalism: The U.S. Environmental Movement, 1970–1990*, ed. Riley E. Dunlap and Angela G. Mertig (Philadelphia: Taylor & Francis, 1992).

18 Christian Joppke, *Mobilizing Against Nuclear Energy: A Comparison of Germany and the United States* (University of California Press, 1993); James M. Jasper and Jane D. Poulsen, "Recruiting Strangers and Friends: Moral Shocks and Social Networks in Animal Rights and Anti-Nuclear Protests," *Social Problems* 42, no. 4 (1995): 493–512.

19 Herbert Gottweis, *Governing Molecules: The Discursive Politics of Genetic Engineering in Europe and the United States* (Cambridge, MA: MIT Press, 1998); Rachel Schurman and William A. Munro, *Fighting for the Future of Food: Activists versus Agribusiness in the Struggle over Biotechnology* (Minneapolis: University of Minnesota Press, 2007); Sheila Jasanoff, *Designs on Nature: Science and Democracy in Europe and the United States* (Princeton, NJ: Princeton University Press, 2005); Abby Kinchy, *Seeds, Science, and Struggle: The Global Politics of Transgenic Crops* (Cambridge, MA: MIT Press, 2012).

20 Sheryl Burt Ruzek, *The Women's Health Movement: Feminist Alternatives to Medical Control* (New York: Praeger, 1978); Steven Epstein, *Impure Science: AIDS, Activism, and the Politics of Knowledge* (Berkeley: University of California Press, 1996); Sabrina McCormick, Julia Brody, Phil Brown, and Ruth Polk "Public Involvement in Breast Cancer Research: An Analysis and Model for Future Research," *International Journal of Health Services* 34, no. 4 (2004): 625–46; Madeleine Akrich, João Nunes, Florence Paterson, and Vololona Rabeharisoa, eds. *The Dynamics of Patient Organizations in Europe* (Paris: Presses de Mines, 2008).

21 National Breast Cancer Coalition, "Project Lead: An Innovative Science Training Program for Breast Cancer Activists," (Washington, DC: National Breast Cancer Coalition Fund, 1998; brochure); Yvonne Andejeski, Isabelle T. Bisceglio, Kay Dickersin, et al.,

"Quantitative Impact of Including Consumers in the Scientific Review of Breast Cancer Research Proposals," *Journal of Women's Health & Gender-Based Medicine* 11, no. 4 (2002): 379–88; Michel Callon and Vololona Rabeharisoa, "Research 'in the Wild' and the Shaping of New Social Identities," *Technology in Society* 25, no. 2 (2003): 193–204; Lawrence W. Green and Shawna L. Mercer, "Can Public Health Researchers and Agencies Reconcile the Push from Funding Bodies and the Pull from Communities?" *American Journal of Public Health* 91, no. 12 (2001): 1926–29.

22 A. M. Chakrabarty, "Patenting of Life-Forms: From a Concept to Reality," in *Who Owns Life?* ed. David Magnus and Arthur Caplan (Amherst, NY: Prometheus Books, 2002).

23 Brief for the People's Business Commission as Amicus Curiae, Diamond v. Chakrabarty, 447 U.S. 303 (1980) (No. 79-136).

24 Interview with former staff of People's Business Commission, July 26, 2011.

25 Jelena Karanović, "Contentious Europeanization: The Paradox of Becoming European Through Anti-Patent Activism," *Ethnos* 75, no. 3 (2010): 252–74; Kelsi Bracmort and Richard K. Lattanzio, "Geoengineering: Governance and Technology Policy," *Congressional Research Service*, November 26, 2013. Online at https://www.fas.org/sgp/crs/misc/R41371.pdf.

26 Frederik Miegel and Tobias Olsson, "From Pirates to Politicians: The Story of the Swedish File Sharers Who Became a Political Party," in *Democracy, Journalism, and Technology: New Developments in an Enlarged Europe.* Online at http://iamcr.org/democracy-journalism-and-technology; Christopher May and Susan K. Sell, *Intellectual Property Rights: A Critical History* (Boulder, CO: Lynne Rienner, 2005).

27 Brigitte van Beuzekom and Anthony Arundel, "OECD Biotechnology Statistics 2009." Online at http://www.oecd.org/sti/42833898.pdf; World Intellectual Property Organization, "World Intellectual Property Indicators," 2014 Report. Online at http://www.wipo.int/edocs/pubdocs/en/wipo_pub_941_2014.pdf.

28 Carlos Correa, *Integrating Public Health Concerns into Patent Legislation in Developing Countries* (Geneva, Switzerland: South Centre, 2000); Stephanie T. Rosenberg, "Asserting the Primacy of Health Over Patent Rights: A Comparative Study of the Processes That Led to the Use of Compulsory Licensing in Thailand and Brazil," *Developing World Bioethics* 14, no. 2 (2014): 83–91.

29 Matthew Rimmer, "Myriad Genetics: Patent Law and Genetic Testing," *European Intellectual Property Review* 25, no. 1 (2003): 20–33; Barbara A. Brenner, "Our Genes Belong to Us. For Now." Online at http://womensfoundationofcalifornia.com/2010/04/08/our-genes-belong-to-us-for-now/; American Civil Liberties Union, "BRCA—Plaintiff Statements." Online at https://www.aclu.org/brca-plaintiff-statements?redirect=free-speech_womens-rights/brca-plaintiff-statements#matloff;Institut Curie, "Against Myriad Genetics's monopoly on tests for predisposition to breast and ovarian cancer, the Institut Curie is initiating an opposition procedure with the European Patent Office" (press release, September 12, 2001), on file with author.

30 Rebecca Skloot, *The Immortal Life of Henrietta Lacks* (New York: Random House, 2010); Universities Allied for Essential Medicines, "Universities Should Do More to Advance Biomedical Research for Neglected Diseases" (press release, April 21, 2015). Online at http://uaem.org/press-release-universities-should-do-more-to-advance-biomedical-research-for-neglected-diseases/; Peter Lee, "Toward a Distributive Commons in Patent Law," *Wisconsin Law Review* 2009, no. 4 (2009): 917–1016.

31 R. K. Gupta and L. Balasubrahmanyam "The Turmeric Effect," *World Patent Information* 20, nos. 3–4 (1998): 185–91; Gerard Bodeker, "Traditional Medical Knowledge, Intellectual Property Rights, and Benefit Sharing," *Cardozo Journal of International and Comparative Law* 11 (2003–2004): 785–814; Vandana Shiva, *Biopiracy: The Plunder of Nature and Knowledge* (Brooklyn, NY: South End Press, 1999).

32 Elon Musk, "All Our Patent Are Belong to You," TESLA (blog), June 12, 2014. Online at http://www.teslamotors.com/blog/all-our-patent-are-belong-you.

33 Arti Rai and James Boyle, "Synthetic Biology: Caught between Property Rights, the Public Domain, and the Commons," PLOS *Biology* 5, no. 3 (2007): 389–93; Michael Heller and Rebecca Eisenberg, "Can Patents Deter Innovation?" *Science* 280 (1998): 698–701.

34 *Transgenic Animal Patent Reform Act of 1989: Hearing on H.R. 1556 Before the H. Subcomm. on Courts, Intellectual Property, and the Administration of Justice of the Comm. on the Judiciary*, 101st Cong. (1989).

35 Jaydee Hanson and Eric Hoffman, "Synthetic Biology: The Next Wave of Patents on Life," in *GeneWatch*. Online at http://www.councilforresponsiblegenetics.org/genewatch/Gene WatchPage.aspx?pageId=310.

36 US Patent and Trademark Office, *Report on Confirmatory Diagnostic Test Activity*, Report to Congress, September 2015.

37 Pompidou, Alain, G2/06 Comments by the President of the European Patent Office, EP 96903521.1-2401/0770125, EPO Register Plus database, on file with author; European Patent Office, "Scenarios for the Future," 2007 Report. Online at http://documents.epo .org/projects/babylon/eponet.nsf/0/63A726D28B589B5BC12572DB00597683/SFile /EPO_scenarios_bookmarked.pdf, 2.

38 Jasanoff, *Designs on Nature*.

39 Lyman Tower Sargent, *Contemporary Political Ideologies: A Comparative Analysis* (Belmont, CA: Wadsworth, 2009).

40 R. Carl Moy, "The History of the Patent Harmonization Treaty: Economic Self-Interest as an Influence," *John Marshall Law Review* 26 (1993): 457–95.

41 "Inside WIPO," WIPO 2014 website. Online at http://www.wipo.int/about-wipo/en/.

42 The third member of the Trilateral Co-operation is the Japan Patent Office. See "About," on the Trilateral 2015 website. Online at http://www.trilateral.net/about.html.

43 America Invents Act of 2011, 35 U.S.C. §112–29.

44 Strathern, "Potential Property, Intellectual Rights, and Property in Persons"; Peter Drahos, *The Global Governance of Knowledge: Patent Offices and Their Clients* (New York: Cambridge University Press, 2010).

45 "Guidelines for Examination in the European Patent Office," European Patent Office Web site. Online at https://www.epo.org/law-practice/legal-texts/guidelines.html.

46 Ibid.

47 World Intellectual Property Organization, IP Statistics Data Center, 2016. Online at http://ipstats.wipo.int/ipstatv2/index.htm?tab=patent.

48 John F. Duffy, "The FCC and the Patent System: Progressive Ideals, Jacksonian Realism, and the Technology of Regulation," *University of Colorado Law Review* 71 (2000): 1071.

49 Susana Borrás, "The Governance of the European Patent System: Effective and Legitimate?" *Economy and Society* 35, no. 4 (2007): 594–610.

50 "Unitary Patent," European Patent Office Web site. Online at https://www.epo.org/law -practice/unitary/unitary-patent.html.

51 "European patents and patent statistics," European Patent Office Web site. Online at
http://www.epo.org/about-us/annual-reports-statistics/statistics.html.

52 World Intellectual Property Organization, IP Statistics Data Center, 2016. Online at
http://ipstats.wipo.int/ipstatv2/index.htm?tab=patent.

53 Jack Walker, *Mobilizing Interest Groups in America: Patrons, Professions, and Social Movements* (Ann Arbor: University of Michigan Press, 1991); Jonathan P. Doh and Terrence R.
Guay, "Corporate Social Responsibility, Public Policy, and NGO Activism in Europe and
the United States: An Institutional-Stakeholder Perspective," *Journal of Management
Studies* 43, no. 1 (2006): 47–73.

54 David Vogel, *The Politics of Precaution: Regulating Health, Safety, and Environmental Risks
in Europe and the United States* (Princeton, NJ: Princeton University Press, 2012).

55 Gøsta Esping-Andersen, ed., *Welfare States in Transition: National Adaptations in Global
Economies* (Thousand Oaks, CA: Sage, 1996); Arnold Heidenheimer, Hugh Heclo, and
Carolyn Teich Adams, *Comparative Public Policy: The Politics of Social Choice in Europe
and America* (New York: St. Martins Press, 1990); Theda Skocpol, *Social Policy in the
United States* (Princeton, NJ: Princeton University Press, 1995); Paul Starr, *The Social
Transformation of American Medicine: The Rise of a Sovereign Profession and the Making of
a Vast Industry* (New York: Basic Books, 1984).

56 Jasanoff, *Designs on Nature*, 255.

57 Marion Fourcade, *Economists and Societies: Discipline and Profession in the United States,
Britain, and France, 1890s to 1990s* (Princeton, NJ: Princeton University Press, 2009);
Arthur Daemmrich, *Pharmacopolitics: Drug Regulation in the United States and Germany*
(Chapel Hill: University of North Carolina Press, 2004).

58 See, for example, Fred Block and Matthew R. Keller, *State of Innovation: The U.S. Government's Role in Technology Development* (Boulder, CO: Paradigm, 2011).

59 Adam Briggle, *A Rich Bioethics: Public Policy, Biotechnology, and the Kass Council* (South
Bend, IN: University of Notre Dame Press, 2010); Susan E. Kelly, "Public Bioethics and
Publics: Consensus, Boundaries, and Participation in Biomedical Science Policy," *Sci-
ence, Technology, and Human Values* 28, no. 3 (2003): 339–64.

60 Schurman and Munro, *Fighting for the Future of Food*.

61 Jasanoff, *Designs on Nature*.

62 Mrill Ingram and Helen Ingram, "Creating Credible Edibles: The Organic Agriculture
Movement and the Emergence of US Federal Organic Standards," in *Routing the Opposition: Social Movements, Public Policy, and Democracy* (Minneapolis: University of Minnesota Press, 2005).

63 Vogel, *Politics of Precaution*; Sheila Jasanoff, *Risk Management and Political Culture*
(Thousand Oaks, CA: Sage, 1986).

64 "Guide to US Regulation of Genetically Modified Food and Agricultural Biotechnology
Products," Pew Initiative on Food and Biotechnology Web site. Online at http://www
.pewtrusts.org/en/research-and-analysis/reports/2001/09/03/guide-to-us-regulation-of
-genetically-modified-food-and-agricultural-biotechnology-products.

65 Les Levidow, Susan Carr, and David Wield, "Genetically Modified Crops in the European
Union: Regulatory Conflicts as Precautionary Opportunities," *Journal of Risk Research* 3,
no. 3 (2000): 189–208.

66 This approach is inspired by work exploring the "co-production" of scientific and political order. Science and technology are mutually constituted, these scholars argue,

with particular social and political orderings, including a society's orienting categories, values, laws, rights, and definitions of knowledge and expertise in policymaking; see Sheila Jasanoff, *States of Knowledge: The Co-production of Science and the Social Order* (New York: Routledge, 2004). Throughout this book, I suggest that *patents* and political order are similarly co-produced in the United States and Europe.

67 Shobita Parthasarathy, "Breaking the Expertise Barrier: Understanding Activist Challenges to Science and Technology Policy Domains," *Science & Public Policy* 37, no. 5 (2010): 355–67.

68 Mirowski, *Science-Mart*; E. P. Berman, *Creating the Market University: How Academic Science Became an Economic Engine* (Princeton, NJ: Princeton University Press, 2011); Adam B. Jaffe and Josh Lerner, *Innovation and Its Discontents: How Our Broken Patent System Is Endangering Innovation and Progress, and What to Do About It* (Princeton, NJ: Princeton University Press, 2011); Michele Boldrin and David K. Levine, *Against Intellectual Monopoly* (New York: Cambridge University Press, 2010); James Boyle, *The Public Domain: Enclosing the Commons of the Mind* (New Haven, CT: Yale University Press, 2010).

CHAPTER ONE

1 Peter Drahos, *The Global Governance of Knowledge: Patent Offices and Their Clients* (New York: Cambridge University Press, 2010).

2 Christopher May and Susan K. Sell, *Intellectual Property Rights: A Critical History* (Boulder, CO: Lynne Reinner, 2005).

3 Graham Dutfield, "Patent Systems as Regulatory Institutions," *Indian Economic Journal* 54, no. 1 (2006): 62–90; Carolyn Deere, *The Implementation Game: The TRIPS Agreement and the Global Politics of Intellectual Property Reform in Developing Countries* (New York: Oxford University Press, 2009).

4 Christine MacLeod, *Inventing the Industrial Revolution: The English Patent System, 1660–1800* (New York: Cambridge University Press, 1988); Ikechi Mgbeoji, "The Juridical Origins of the International Patent System: Towards a Historiography of the Role of Patents in Industrialization," *Journal of the History of International Law* 5, no. 2 (2003): 403–22.

5 Edith T. Penrose, *The Economics of the International Patent System* (Baltimore, MD: Johns Hopkins University Press, 1951).

6 Ibid.; William L. Letwin, "The English Common Law Concerning Monopolies," *University of Chicago Law Review* 21, no. 3 (1954): 355–85; Statute of Monopolies (1623) 21 Jac. 1, c. 3 (Eng.).

7 Klaus Boehm, *The British Patent System I: Administration* (New York: Cambridge University Press, 1967).

8 Chris Dent, "'Generally Inconvenient': The 1624 Statute of Monopolies as Political Compromise," *Melbourne University Law Review* 33, no. 2 (2009): 16–56.

9 Liliane Hilarie-Perez, "Invention and the State in 18th-Century France," *Technology and Culture* 32, no. 4 (1991): 911–31; Christine MacLeod, "The Paradoxes of Patenting: Invention and Its Diffusion in 18th- and 19th-Century Britain, France, and North America," *Technology and Culture* 32, no. 4 (1991): 885–910.

10 Gerhart Husserl, "Public Policy and Ordre Public," *Virginia Law Review* 25, no. 1 (1938): 37–67.

11 M. Forde, "The 'Ordre Public' Exception and Adjudicative Jurisdiction Conventions," *International and Comparative Law Quarterly* 29, nos. 2–3 (1980): 259–73.

12 Maître J. B. Bernier, "Droit Public and Ordre Public," *Transactions of the Grotius Society* 15 (1929): 83–91.

13 Boehm, *British Patent System*, vol. 1; Maurice Cassier, "Patents and Public Health in France: Exclusion of Drug Patents and Granting of Pharmaceutical Process Patents between the Two World Wars," *History and Technology* 24, no. 2 (2008): 153–71; Penrose, *Economics of the International Patent System*.

14 Michele Boldrin and David K. Levine, *Against Intellectual Monopoly* (New York: Cambridge University Press, 2010).

15 Committee of Experts on Patents, "Criteria of Novelty and Patentability," Council of Europe, CM/WP IV 9 C53 (1951): 19, on file with author.

16 Boehm, *British Patent System*.

17 Penrose, *Economics of the International Patent System*.

18 Fritz Machlup and Edith T. Penrose, "The Patent Controversy in the Nineteenth Century," *Journal of Economic History* 10, no. 1 (1950): 1–29; Penrose, *Economics of the International Patent System*.

19 Kathryn Steen, *The American Synthetic Organic Chemicals Industry: War and Politics, 1910–1930* (Chapel Hill: University of North Carolina Press, 2014).

20 Penrose, *Economics of the International Patent System*.

21 Paul S. Haar, "Revision of the Paris Convention: A Realignment of Private and Public Interests in the International Patent System," *Brooklyn International Law Journal* 8, no. 1 (1982): 77–108.

22 Esther van Zimmeren and Geertrui Van Overwalle, "A Paper Tiger? Compulsory License Regimes for Public Health in Europe," *International Review of Intellectual Property and Competition Law (IIC)* (December 1, 2010): 1–37. Online at http://www.researchgate.net /publication/228265672_A_Paper_Tiger_Compulsory_License_Regimes_for_Public _Health_in_Europe/file/9fcfd50ed84a6a4b18.pdf; Adrian Johns, "Intellectual Property and the Nature of Science," *Cultural Studies* 20, nos. 2–3 (2006): 145–64.

23 Edmund J. Sease, "Common Sense, Nonsense and the Compulsory License," *Journal of the Patent Office Society* 55, no. 4 (1973): 233–54; van Zimmeren and Van Overwalle, "A Paper Tiger?".

24 Adam Mossoff, "Who Cares What Thomas Jefferson Thought About Patents? Reevaluating the Patent 'Privilege' in Historical Context," *Cornell Law Review* 92 (2007): 953; Edward C. Walterscheid, "Inherent or Created Rights: Early Views on the Intellectual Property Clause," *Hamline Law Review* 19, no. 1 (1995): 81.

25 U.S. Const. art. I, § 8, cl. 8.

26 Patent Act of 1790, Ch. 7, 1 Stat. 109–112 (April 10, 1790).

27 Ibid.

28 Courtney Fullilove, "The Archive of Useful Knowledge" (PhD diss., Columbia University, 2009), ProQuest Dissertations and Theses database. Online at http://search.proquest .com/dissertations/docview/205428195/fulltextPDF/C5E8EF2965034197PQ/1?account id=14667.

29 Mario Biagioli, "Patent Specification and Political Representation: How Patents Became Rights," in *Making and Unmaking Intellectual Property: Creative Production in Legal and Cultural Perspective*, ed. Mario Biagioli, Peter Jaszi, and Martha Woodmansee (Chicago: University of Chicago Press, 2011).

30 Patent Act of 1836, Ch. 357, 5 Stat. 117 (July 4, 1836).

31 Patent Act of 1790, Ch. 7, 1 Stat. 109–112 (April 10, 1790); Patent Act of 1836, Ch. 357, 5 Stat. 117 (July 4, 1836).

32 Senator John Ruggles, Senate Committee on Patents, "The select committee appointed to examine and report the extent of the loss sustained by the burning of the Patent Office," 24th Cong., No. 107 at 58 (January 9, 1837).

33 Kenneth W. Dobyns, *The Patent Office Pony: A History of the Early Patent Office* (Fredericksburg, VA: Sergeant Kirkland's Museum and Historical Society, 1994).

34 "Congress and Inventors," *Scientific American*, April 29, 1848, 253.

35 Roger Hahn, *The Anatomy of a Scientific Institution: The Paris Academy of Sciences, 1666–1803* (Berkeley: University of California Press, 1971).

36 Hilaire-Pérez, "Invention and the State in 18th-Century France"; MacLeod, "Paradoxes of Patenting."

37 Robert C. Post, "'Liberalizers' versus 'Scientific Men' in the Antebellum Patent Office," *Technology and Culture* 17, no. 1 (1976): 24–54.

38 Kara W. Swanson, "The Emergence of the Professional Patent Practitioner," *Technology and Culture* 50, no. 3 (2009): 519–48.

39 The United States also had a registration system for a brief period, when its initial examination period seemed too onerous. It instituted the system in its 1793 Patent Act but ended it (and replaced it with a new, bureaucratic approach to examination) with the 1836 Patent Act.

40 Swanson, "Emergence of the Professional Patent Practitioner."

41 Ibid.

42 "US Patent Activity, Calendar Years 1790 to the Present," US Patent and Trademark Office Website. Online at http://www.uspto.gov/web/offices/ac/ido/oeip/taf/h_counts.htm.

43 Dobyns, *Patent Office Pony*.

44 Yaron Ezrahi, *The Descent of Icarus: Science and the Transformation of Contemporary Democracy* (Cambridge, MA: Harvard University Press, 1990); Sheila Jasanoff, *The Fifth Branch: Science Advisers as Policymakers* (Cambridge, MA: Harvard University Press, 1990).

45 In interviews, PTO personnel stated repeatedly that they interpreted their commitment to the public interest in terms of procedural objectivity. One supervisory examiner stated that she felt her goal was to "keep an open claim field for all comers to come and be able to [be rewarded for their inventions] and to share, because that's how we move forward as a country. . . . I think for us to keep it fair and keep it even for everybody, that's how we serve the public interest." Interview with PTO Employee C, April 3 2009. Another supervisory examiner defended the PTO by arguing that the patents that have been challenged were assessed appropriately: "When people are angry at us, usually I look at the work of my examiners. I'm quite proud of the work they do. Usually we're making rational decisions on the evidence that's in front of us." (PTO Employee D 2009) Many current and former PTO personnel also call attention to the involvement of quality-control specialists, who review patent decisions at random to ensure that rules are being followed to minimize individual judgments. Interview with PTO Employee C, April 3 2009; Interview with PTO Employee E, January 18, 2008.

46 Margo A. Bagley, "Patent First, Ask Questions Later: Morality and Biotechnology in Patent Law," *William and Mary Law Review* 45, no. 2 (2003): 469–547; Margo A. Bagley, "A Global Controversy: The Role of Morality in Biotechnology Patent Law," University of

Virginia Law School Public Law and Legal Theory Working Paper Series, Paper 57 (2007); Cynthia M. Ho, "Splicing Morality and Patent Law: Issues Arising from Mixing Mice and Men," *Washington University Journal of Law & Policy* 2 (2000): 247–85.

47 Lowell v. Lewis, 1 Mason. 182; 1 Robb, Pat. Cas. 131 (Circuit Court, D. Massachusetts. May Term. 1817).

48 Alfred E. Kahn, "Fundamental Deficiencies of the American Patent Law," *American Economic Review* 30, no. 3 (1940): 475–91.

49 Benjamin D. Enerson, "Protecting Society from Patently Offensive Inventions: The Risk of Reviving the Moral Utility Doctrine," *Cornell Law Review* 89 (2004): 685; Andrew R. Smith, "Monsters at the Patent Office: The Inconsistent Conclusions of Moral Utility and the Controversy of Human Cloning," *DePaul Law Review* 53, no. 1 (2003): 159.

50 Nathan Machin, "Prospective Utility: A New Interpretation of the Utility Requirement of Section 101 of the Patent Act," *California Law Review* 87, no. 2 (1999): 421–56; Laura A. Keay, "Morality's Move Within U.S. Patent Law: From Moral Utility to Subject Matter," *AIPLA Quarterly Journal* 40, no. 3 (2012): 409–39.

51 Joe Gabriel, *Medical Monopoly: Intellectual Property Rights and the Origins of the Modern Pharmaceutical Industry* (Chicago: University of Chicago Press, 2014).

52 Ibid.

53 *Oldfield Revision and Codification of the Patent Laws: Hearing Before the H. Committee of Patents*, 63rd Cong. 1–174 (1914).

54 Stefan A. Risenfeld, "Licenses and United States Industrial and Artistic Property Law," *California Law Review* 47, no. 1 (1959): 51–63.

55 *Oldfield Revision and Codification of the Patent Laws: Hearing on H.R. 23417 Before the H. Committee of Patents*, 62nd Cong. 8 (1912).

56 *Oldfield Revision and Codification of the Patent Laws: Hearing Before the H. Committee of Patents*, 63rd Cong. 1–19 (1914) (statement of Mr. Redding, Merchants' Association, New York City).

57 *Oldfield Revision and Codification of the Patent Laws: Hearing Before the H. Committee of Patents*, 63rd Cong. 24 (1914) (statement of Mr. William W. Dodge, Patent Law Association, Washington, DC).

58 Ibid., 28.

59 Steen, *American Synthetic Organic Chemicals Industry*.

60 *Salvarsan: Hearing on S. 2178 and 2363 Before the S. Committee on Patents*, 65th Cong. 5 (1917).

61 J. M. T. Finney and George Walker, "Abrogate the Patent on Salvarsan," *Journal of the American Medical Association* 68, no. 21 (1917): 1573.

62 Steen, *American Synthetic Organic Chemicals Industry*.

63 *Salvarsan: Hearing on S. 2178 and 2363 Before the S. Committee on Patents*, 65th Cong. 5 (1917).

64 *Revision of Statues Relating to Patents: Hearing on S. 3325 and S. 3410 Before the S. Committee Patents*, 67th Cong. 7 (1922).

65 Ibid., 202.

66 Susan Sell and Christopher May, "Moments in Law: Contestation and Settlement in the History of Intellectual Property," *Review of International Political Economy* 8, no. 3 (2001): 467–500.

67 Ibid.

68 Jeremy Greene and Scott H. Podolsky, "Reform, Regulation, and Pharmaceuticals—The Kefauver-Harris Amendments at 50," *New England Journal of Medicine* 367 (2012): 1481–83.

69 *Drug Industry Antitrust Act: Hearing on S. 1552 and S. 6245 Before the S. Committee on Antitrust and Monopoly*, 87th Cong. 1192 (1961).

70 Richard Harris, *The Real Voice* (New York: Macmillan, 1964).

71 Ibid.

72 *Drug Industry Antitrust Act: Hearing on S. 1552 Before the S. Subcommittee on Antitrust and Monopoly of the Committee on the Judiciary*, 87th Cong. 1191–1989 (1961); Dominique Tobbell, *Pills, Power, and Policy: The Struggle for Drug Reform in Cold War America and Its Consequences* (Berkeley: University of California Press, 2012).

73 *Drug Industry Antitrust Act: Hearing on S. 1552 Before the S. Subcommittee on Antitrust and Monopoly of the Committee on the Judiciary*, 87th Cong. 1474 (1961).

74 Ibid., 1477.

75 Penrose, *Economics of the International Patent System*; Robert Van Horn and Mattias Klaes, "Chicago Neoliberalism versus Cowles Planning," *Journal of the History of the Behavioral Sciences* 47, no. 3 (2011): 302–21.

76 Fritz Machlup, *An Economic Review of the Patent System* (Washington, DC: US Government Printing Office, 1958).

77 Machlup and Penrose, "Patent Controversy in the Nineteenth Century."

78 *Drug Industry Antitrust Act: Hearing on S. 1552 Before the S. Subcommittee on Antitrust and Monopoly of the Committee on the Judiciary*, 87th Cong. 1384 (1961).

79 Penrose, *Economics of the International Patent System*.

80 Marion Fourcade, *Economists and Societies: Discipline and Profession in the United States, Britain, and France, 1890s to 1990s* (Princeton, NJ: Princeton University Press, 2009); Theodore M. Porter, *Trust in Numbers: The Pursuit of Objectivity in Science and Public Life* (Princeton, NJ: Princeton University Press, 1996).

81 *Drug Industry Antitrust Act: Hearing on S. 1552 Before the S. Subcommittee on Antitrust and Monopoly of the Committee on the Judiciary*, 87th Cong. 1393 (1961).

82 Joseph Bruce Gorman, *Kefauver: A Political Biography* (New York: Oxford University Press, 1971).

83 Simone Turchetti, "Patenting the Atom: The Controversial Management of State Secrecy and Intellectual Property Rights in Atomic Research," in *Knowledge Management and Intellectual Property: Concepts, Actors, and Practices from the Past to the Present*, ed. Stathis Arapostathis and Graham Dutfield (Northampton, MA: Edward Elgar, 2013); Steen, *American Synthetic Organic Chemicals Industry*.

84 See, for example, Hilarie-Perez, "Invention and the State in 18th-Century France"; Richard Spencer, "The German Patent Office," *Journal of the Patent Office Society* 31, no. 2 (1949): 79–87; B. Zorina Khan (n.d.), "An Economic History of Patent Institutions." Online at http://eh.net/encyclopedia/an-economic-history-of-patent-institutions/.

85 Hans Shade, *Patents at a Glance: A Survey of Substantive Law and Formalities in 46 Countries* (Munich: Carl Heymanns Verlag KG, 1971).

86 Committee of Experts on Patents, "Criteria of Novelty and Patentability," 2.

87 Drahos, *Global Governance of Knowledge*.

88 Dominique Guellec and Bruno van Pottelsberghe de la Potterie, *The Economics of the European Patent System: IP Policy for Innovation and Competition* (New York: Oxford Uni-

versity Press, 2007); Eda Kranakis, "Patents and Power: European Patent-System Integration in the Context of Globalization," *Technology and Culture* 48 (2007): 689–728.

89 Statute of the Council of Europe, Council of Europe, May 5, 1949, London, 5.V. 1949. Online at http://conventions.coe.int/treaty/en/Treaties/Html/001.htm; Sidney Pollard, *The Integration of the European Economy Since 1815* (New York: Routledge, 2006).

90 Committee on Legal and Administrative Questions, "European Patents Office," Council of Europe, Recommendation 23, Doc 110 (1949), on file with author.

91 Written Question No. 73 to the Committee of Ministers, Council of Europe, on file with author.

92 Committee of Experts on Patents, "Criteria of Novelty and Patentability," Council of Europe, CM/WP IV (51) 9 (1951), on file with author.

93 Ibid., 2.

94 Ibid.

95 Ibid.

96 Machlup and Penrose, "Patent Controversy in the Nineteenth Century."

97 Committee of Experts on Patents, Council of Europe, "Report of the Committee of Experts to the Committee of Ministers on the meeting held at Strasbourg from 14th to 17th May 1963: Annex II (Declaration by the Austrian delegation)," CM (63) 101 (27 May 1963) 16, on file with author.

98 Lodovico Benvenuti, "Written Question No. 73 to the Committee of Ministers by Mr. Heckscher (CM/AS [60]Quest73FinalE," [1960]), on file with author.

99 Committee of Experts on Patents, Council of Europe, "Observations of the Committee of National Institutes of Patent Agents (CNIPA) on the two draft Conventions being prepared by the Committee of Experts," EXP/Brev (62) 6 (June 28, 1962): 4, on file with author.

100 Committee of Experts on Patents, Council of Europe, "The Patentability of 'Products,'" CM/WP IV (51) 27 (1951): 3, on file with author.

101 C. Wadlow, "Strasbourg, the Forgotten Patent Convention, and the Origins of the European Patents Jurisdiction," *International Review of Intellectual Property and Competition Law* 41, no. 2 (2010): 123–49.

102 Keith Aoki, *Seed Wars: Cases and Materials on Intellectual Property and Plant Genetic Resources* (Durham, NC: Carolina Academic Press, 2008).

103 Ibid.

104 European Patent Office, "Revision of the European Patent Convention (EPC 2000) Synoptic Presentation EPC 1973/2000—Part I: The Articles," European Patent Office, Munich, Germany (2007): 50, on file with author.

105 Oksana Mitnovetski and Dianne Nicol, "Are Patents for Methods of Medical Treatment Contrary to the Ordre Public and Morality or 'Generally Inconvenient'?" *Journal of Medical Ethics* 30 (2004): 470–75; S. Tina Piper, "A Common Law Prescription for a Medical Malaise," in *The Common Law of Intellectual Property: Essays in Honour of Professor David Vaver*, ed. C. Ng, L. Bently and G. D'Agostino (Oxford: Hart Publishing, 2010); Stefan Bechtold, "Physicians as User Innovators," in *Intellectual Property at the Edge: The Contested Contours of IP*, ed. Rochelle C. Dreyfuss and Jane C. Ginsburg (New York: Cambridge University Press, 2014); Dominique Ritter, "Switzerland's Patent Law History," *Fordham Intellectual Property, Media, and Entertainment Law Journal* 14 (2004): 463–96.

106 Dennis Thompson, "The Draft Convention for a European Patent," *International and*

Comparative Law Quarterly 22, no. 1 (1973): 51–82; Franz Froschmaier, "Some Aspects of the Draft Convention Relating to a European Patent Law," *International and Comparative Law Quarterly* 12, no. 3 (1963): 886–97.

107 Joseph Weiler, *The Constitution of Europe: "Do the New Clothes Have an Emperor?" and Other Essays on European Integration* (New York: Cambridge University Press, 1999).

108 Thompson, "Draft Convention for a European Patent."

109 Ibid.

110 The America Invents Act, passed in 2011, includes for the first time a post-grant opposition mechanism, but it is more narrowly drawn than its European counterpart. Third parties can only "oppose" a patent on the basis of an error in the review of prior art. See "The Opposition Procedure." Online at the European Patent Office Web site, last modified September 9, 2008, http://www.epo.org/about-us/jobs/examiners/what/opposition.html.

111 Stuart J. H. Graham, Bronwyn H. Hall, Dietmar Harhoff, and David C. Mowery, "Post-Issue Patent 'Quality Control': A Comparative Study of US Patent Re-Examinations and European Patent Oppositions," National Bureau of Economic Research Working Paper Series 8807 (February 2002). Online at http://www.nber.org/papers/w8807.

112 European Patent Office, "Engineers and Scientists: Interested in Patent Work?" *The Observer* (UK), July 7, 1973, 33.

113 "Objectives," Trilateral. Online at http://www.trilateral.net/about/objectives.html.

CHAPTER TWO

1 Susan Wright, *Molecular Politics: Developing American and British Regulatory Policy for Genetic Engineering, 1972–1982* (Chicago: University of Chicago Press, 1994); Clyde A. Hutchison, "DNA Sequencing: Bench to Bedside and Beyond," *Nucleic Acids Research* 35 (2007): 6227–37.

2 Wright, *Molecular Politics*; David L. Teichmann, "Regulation of Recombinant DNA Research: A Comparative Study," *Loyola of Los Angeles International and Comparative Law Review* 6, no. 1 (1983): 1–35.

3 Wright, *Molecular Politics*; Teichmann, "Regulation of Recombinant DNA Research."

4 Nicholas Rasmussen, *Gene Jockeys: Life Science and the Rise of Biotech Enterprise* (Baltimore, MD: Johns Hopkins University Press, 2014).

5 Sally S. Hughes, "Making Dollars Out of DNA: The First Major Patent in Biotechnology and the Commercialization of Molecular Biology, 1974–1980," *Isis* 92, no. 3 (2001): 541–75.

6 Patent Act of 1952, 35 U.S.C. §§ 1–318 (1952).

7 George W. Mazon III, "Products of Nature: The New Criteria," *Catholic University Law Review* 20, no. 4 (1971): 783–90.

8 Daniel J. Kevles, "Ananda Chakrabarty Wins a Patent: Biotechnology, Law, and Society," *Historical Studies in the Physical and Biological Sciences* 25, no. 1 (1994): 111–35.

9 Ibid.

10 Sheila Jasanoff, "Taking Life: Private Rights in Public Nature," in *Lively Capital: Biotechnologies, Ethics, and Governance in Global Markets*, ed. Kaushik Sunder Rajan (Durham, NC: Duke University Press, 2012): 155–83.

11 Hughes, "Making Dollars Out of DNA."

12 Cary Fowler, "The Plant Patent Act of 1930: A Sociological History of Its Creation," *Journal of the Patent and Trademark Office Society* 82 (2000): 621–44.

13 To draw this conclusion, I analyzed all of the amicus briefs submitted to the Supreme Court in patent-related cases throughout its history. Patent-law associations and large companies submitted the most amicus briefs, and these participants were also most prominently featured in congressional hearings. (I draw this conclusion by evaluating all patent-related Congressional hearings since the 18th century.) This finding is validated by the following article: Colleen V. Chien, "Patent Amicus Briefs: What the Courts' Friends Can Teach Us about the Patent System," Santa Clara University School of Law Legal Studies Research Paper Series, Paper No. 10-05 (2010): 1–28. Online at http://ssrn.com/abstract=1608111.

14 Diamond v. Chakrabarty, 447 U.S. 303 (1980) (No. 79-136): Brief for the People's Business Commission as Amicus Curiae; Brief for the American Society for Microbiology as Amicus Curiae; Brief for the American Patent Law Association, Inc. as Amicus Curiae; Brief for the Pharmaceutical Manufacturers Association as Amicus Curiae; Brief for the Regents of the University of California as Amicus Curiae; Brief for Genentech as Amicus Curiae; Brief for the New York Patent Law Association as Amicus Curiae; Brief for the Dr. Leroy E. Hood, Dr. Thomas P. Maniatis, Dr. David S. Eisenberg, The American Society Of Biological Chemists, The Association Of American Medical Colleges, The California Institute of Technology, The American Council On Education as Amicus Curiae; Brief for Dr. George Pieczenik as Amicus Curiae.

15 James F. Spriggs and Paul J. Wahlbeck, "Amicus Curiae and the Role of Information at the Supreme Court," *Political Research Quarterly* 50, no. 2 (1997): 365–86.

16 Brief for the American Patent Law Association, at 17.

17 Brief for the New York Patent Law Association.

18 Fowler, "Plant Patent Act of 1930," 621.

19 Brief for the American Patent Law Association, at 16.

20 Brief for Genentech, at 17.

21 Brief for the Pharmaceutical Manufacturers Association, at 24.

22 Rachel Schurman and William A. Munro, *Fighting for the Future of Food: Activists versus Agribusiness in the Struggle over Biotechnology* (Minneapolis: University of Minnesota Press, 2010); Herbert Gottweis, *Governing Molecules: The Discursive Politics of Genetic Engineering in Europe and the United States* (Cambridge, MA: MIT Press, 1998); R. Stephen Crespi, "The European Biotechnology Directive Is dead," *Trends in Biotechnology* 13, no. 5 (1995): 162–64; Sheila Jasanoff, *Designs on Nature: Science and Democracy in Europe and the United States* (Princeton, NJ: Princeton University Press, 2005).

23 Paul S. Naik, "Biotechnology Through the Eyes of An Opponent: The Resistance of Activist Jeremy Rifkin," *Virginia Journal of Law and Technology* 5, no. 2 (2000): 5–15.

24 Ted Howard, *The P.B.C.: A History* (Washington, DC: The People's Bicentennial Commission, 1976).

25 Christopher B. Daly, "The People's Bicentennial Commission: Slouching Towards the Economic Revolution," *Harvard Crimson*, April 28, 1975. Online at http://www.thecrimson.com/article/1975/4/28/the-peoples-bicentennial-commission-pif-you/; "Jeremy Rifkin." Online at http://www.knology.net/~bilrum/rifkin.htm.

26 See, for example, Jeremy Rifkin, *Own Your Own Job: Economic Democracy for Working*

Americans (New York: Bantam Books, 1977); People's Bicentennial Commission, *Common Sense II: The Case Against Corporate Tyranny* (New York: Bantam Books, 1975); Jeremy Rifkin and Randy Barber, *The North Will Rise Again: Pensions, Politics, and Power in the 1980s* (Boston: Beacon Press, 1978).

27 Kelly Moore, *Disrupting Science: Social Movements, American Scientists, and the Politics of the Military, 1945–1975* (Princeton, NJ: Princeton University Press, 2008).

28 This close connection is clear from the acknowledgements section of Howard and Rifkin's first major publication on biotechnology, *Who Should Play God? The Artificial Creation of Life and What It Means for the Future of the Human Race.* The authors acknowledge the help of StfP, which they describe as "a nationwide organization of scientists who have devoted their lives to insuring that science and technology serve humanity rather than enslave it." Ted Howard and Jeremy Rifkin, *Who Should Play God?: The Artificial Creation of Life and What It Means for the Future of the Human Race* (New York: Dell, 1977), 4.

29 "DNA Law Likely," *Chemical Week* (1977): 26.

30 Howard and Rifkin, *Who Should Play God?*

31 Moore, *Disrupting Science.*

32 Howard and Rifkin, *Who Should Play God?* 8.

33 Ibid., 203.

34 Interview with former staff of People's Business Commission, July 26, 2011.

35 Naik, "Biotechnology Through the Eyes of An Opponent."

36 Ibid.

37 Fowler, "Plant Patent Act of 1930."

38 Brief for the People's Business Commission, at 10.

39 Ibid.

40 Ibid., 29, 31.

41 Ibid., 25.

42 Ibid.

43 Ibid.

44 "The Patented Life," *New York Times*, June 18, 1980, A30; Wright, *Molecular Politics.*

45 Ibid.

46 Brief for the American Patent Law Association, at 21.

47 Brief for the Pharmaceutical Manufacturers Association, at i.

48 Diamond v. Chakrabarty, 447 U.S. 303 (1980), 5.

49 Diamond v. Chakrabarty, 447 U.S. 303 (1980), 12.

50 Diamond v. Chakrabarty, 447 U.S. 303 (1980), 12.

51 Diamond v. Chakrabarty, 447 U.S. 303 (1980), 4.

52 Diamond v. Chakrabarty, 447 U.S. 303 (1980), 17.

53 "Fresh Debate Over the Life-form Ruling," *Chemical Week*, August 6, 1980, 47.

54 Kieran Healy, *Last Best Gifts: Altruism and the Market for Human Blood and Organs* (Chicago: University of Chicago Press, 2006); Kara Swanson, *Banking on the Body: The Market in Blood, Milk, and Sperm in Modern America* (Cambridge, MA: Harvard University Press, 2014).

55 Elizabeth Popp Berman, *Creating the Market University: How Academic Science Became An Economic Engine* (Princeton, NJ: Princeton University Press, 2012).

56 Philip Mirowski, *Science-Mart: Privatizing American Science* (Cambridge, MA: Harvard University Press, 2011).

57 Derek Bok, *Universities in the Marketplace: The Commercialization of Higher Education* (Princeton, NJ: Princeton University Press, 2004).

58 Philip W. Grubb and Peter R. Thomsen, *Patents for Chemicals, Pharmaceuticals and Biotechnology: Fundamentals of Global Law, Practice, and Strategy*, 5th ed. (New York: Oxford University Press, 2010).

59 E. Richard Gold and Alain Gallochat, "The European Biotech Directive: Past as Prologue," *European Law Journal* 7, no. 3 (2001): 331–66.

60 Ibid.

61 Robert Bud, *The Uses of Life: A History of Biotechnology* (Cambridge: Cambridge University Press, 1994); Gerald Kamstra, Mark Döring, Nick Scott-Ram, Andrew Sheard, and Henry Wixon, *Patents on Biotechnological Inventions: The E. C. Directive* (London: Sweet and Maxwell, 2002).

62 Schurman and Munro, *Fighting for the Future of Food*; Gottweis, *Governing Molecules*; Crespi, "European Biotechnology Directive Is Dead."

63 Crespi, "European Biotechnology Directive Is Dead."

64 Commission Proposal for a Council Directive on the Legal Protection of Biotechnological Inventions, at 5, COM (1988) 496 final (October 17, 1988).

65 Commission Proposal for a Council Directive on the Legal Protection of Biotechnological Inventions, COM (1988) 496 final (October 17, 1988).

66 Commission Proposal for a Council Directive on the Legal Protection of Biotechnological Inventions, at 4, COM (1988) 496 final (October 17, 1988).

67 Bernard Steunenberg, "Playing Different Games: The European Parliament and the Reform of Codecision," in *The European Parliament: Moving toward Democracy in the EU*, ed. Bernard Steunenberg and Jacques Thomassen (New York: Rowman & Littlefield, 2002), 163.

68 *Opinion of the Economic and Social Committee on the 'Proposal for a European Parliament and Council Directive on the Legal Protection of Biotechnological Inventions,'* 1989 O.J. (C 159) 10.

69 Bernard Steunenberg, "Playing Different Games: The European Parliament and the Reform of Codecision," in *The European Parliament: Moving toward Democracy in the EU*, ed. Bernard Steunenberg and Jacques Thomassen (New York: Rowman & Littlefield, 2002), 163.

70 The European Parliament, "Resolution on the Ethical and Legal Problems of Genetic Engineering," *Official Journal C* 96 (1989): 165–171, accessed July 17, 2014, http://www.codex.vr.se/texts/EP-genetic.html#3.

71 European Parliament, Committee on Legal Affairs and Citizens' Rights, *Hearing of experts on the legal protection of biotechnological inventions, Summary Record of Oral Statements*, COM (88) 496 final—SYN 159, May 21 and 22, 1990.

72 "GRAIN-Publications," GRAIN. Online at http://www.grain.org/article/archive/categories/12-publications.

73 *Report of the Committee on Legal Affairs and Citizens' Rights on the Commission Proposal for a Council Directive on the Legal Protection of Biotechnological Inventions*, at 63, COM (88) 496 final (January 29, 1992).

74 Ibid.

75 *Draft Report of the Committee on Development and Cooperation on the Proposal for a Council Directive on the Legal Protection of Biotechnological Inventions*, at 5, COM (88) 496 final (July 3, 1992).

76 Ibid., 4.

77 *Report of the Committee on Legal Affairs and Citizens' Rights on the Commission Proposal for a Council Directive on the Legal Protection of Biotechnological Inventions*, at 48, COM (88) 496 final (January 29, 1992).

78 *Draft Report of the Committee on Agriculture, Fisheries, and Rural Development on the Proposal for a Council Directive on the Legal Protection of Biotechnological Inventions*, at 59, COM (88) 496 final (January 29, 1992).

79 *Draft Report of the Committee on Economic and Monetary Affairs and Industrial Policy on the Proposal for a Council Directive on the Legal Protection of Biotechnological Inventions*, at 6, COM (88) 496 final (January 29, 1992).

80 *Draft Report of the Committee on Agriculture, Fisheries, and Rural Development on the Proposal for a Council Directive on the Legal Protection of Biotechnological Inventions*, at 59, COM (88) 496 final (January 29, 1992).

81 Christopher McCrudden, "Human Dignity and Judicial Interpretation of Human Rights," *European Journal of International Law* 19, no. 4 (2008): 655–724; Stefan Sperling, *Reasons of Conscience: The Bioethics Debate in Germany* (Chicago: University of Chicago Press, 2013); James Q. Whitman, "The Two Western Cultures of Privacy: Dignity versus Liberty," *Yale Law Journal* 113, no. 6 (2004): 1151–1221.

82 *Report of the Committee on Legal Affairs and Citizens' Rights on the Commission Proposal for a Council Directive on the Legal Protection of Biotechnological Inventions*, at 62, COM (88) 496 final (January 29, 1992).

83 Sandra Schmieder, "Scope of Biotechnology Inventions in the United States and in Europe—Compulsory Licensing, Experimental Use and Arbitration: A Study of Patentability of DNA-Related Inventions with Special Emphasis on the Establishment of an Arbitration Based Compulsory Licensing System," *Santa Clara High Technology Law Journal* 21, no. 1 (2004): 164–234; Kevin Iles, "A Comparative Analysis of the Impact of Experimental Use Exemptions in Patent Law on Incentives to Innovate," *Northwestern Journal of Technology and Intellectual Property* 4, no. 1 (2005): 61–82.

84 Willi Rothley, "European Parliament Session Documents," European Parliament Archives, Luxembourg, January 29 (1992), 63, on file with author.

85 Aldo Geuna and Lionel J. J. Nesta, "University Patenting and Its Effects on Academic Research: The Emerging European Evidence," *Research Policy* 35 (2006): 790–807.

86 *Draft Report of the Committee on Legal Affairs and Citizens' Rights on the Proposal for a Council Directive on the Legal Protection of Biotechnological Inventions*, at 25, COM (88) 496 final (May 27, 1991).

87 *Draft Report of the Committee on Legal Affairs and Citizens' Rights on the Proposal for a Council Directive on the Legal Protection of Biotechnological Inventions*, COM (88) 496 final (June 4, 1991).

88 *Report of the Committee on Legal Affairs and Citizens' Rights on the Commission Proposal for a Council Directive on the Legal Protection of Biotechnological Inventions*, at 24, COM (88) 496 final (January 29, 1992).

89 Susan E. Kelly, "Public Bioethics and Publics: Consensus, Boundaries, and Participation in Biomedical Science Policy," *Science, Technology, and Human Values* 28, no. 3 (2003): 339–64; Albert R. Jonsen, *The Birth of Bioethics* (New York: Oxford University Press, 1998); David J. Rothman, *Strangers at the Bedside: A History of How Law and Bioethics Transformed Medical Decision Making* (New York: Basic Books, 1992).

90 Sperling, *Reasons of Conscience*.

91 Helen Busby, Tamara Hervey, and Alison Mohr, "Ethical EU law? The Influence of the European Group on Ethics in Science and New Technologies," *European Law Review* 33 (2008): 803–42.

92 James F. Childress, "Deliberations of the Human Fetal Tissue Transplantation Research Panel," in *Biomedical Politics*, ed. Kathi Hanna (Washington DC: National Academies Press, 1991): 215–57.

93 Michael Mulkay, *The Embryo Research Debate: Science and the Politics of Reproduction* (New York: Cambridge University Press, 1997).

94 Sperling, *Reasons of Conscience*.

95 Hilda Kean, *Animal Rights: Political and Social Change in Britain since 1800*, (London: Reaktion Books, 1998); Kathryn Shevelow, *For the Love of Animals: The Rise of the Animal Protection Movement*, (New York: Holt, 2009).

96 United Kingdom. "Select Committee on Animals in Scientific Procedures Report." The Stationery Office. 2002. Online at http://www.publications.parliament.uk/pa/ld200102 /ldselect/ldanimal/150/15004.htm#a4; Kean, *Animal Rights*.

97 Erin Evans, "Constitutional Inclusion of Animal Rights in Germany and Switzerland: How Did Animal Protection Become an Issue of National Importance?" *Society and Animals* 18 (2010): 231–50.

98 European Commission, "Animals Used for Scientific Purposes," Web site. Online at http://ec.europa.eu/environment/chemicals/lab_animals/nextsteps_en.htm.

99 Diane L. Beers, *For the Prevention of Cruelty: The History and Legacy of Animal Rights Activism in the United States* (Athens, OH: Swallow Press, 2006); James M. Jasper and Dorothy Nelkin, *The Animal Rights Crusade: The Growth of a Moral Protest* (New York: Free Press, 1991).

100 Remarks of Margarida Salema, EUR. PARL. DEB. (3-417) 19 (April 6, 1992).

101 Remarks of Hiltrud Breyer, EUR. PARL. DEB. (3-417) 19 (April 6, 1992).

102 Elizabeth Bomberg, *Green Parties and Politics in the European Union* (New York: Routledge, 1998).

103 Remarks of Alain Pompidou, EUR. PARL. DEB. (3-417) 21 (April 6, 1992).

104 Remarks of Gérard Caudron, EUR. PARL. DEB. (3-423) 304 (April 6, 1992).

105 Crespi, "European Biotechnology Directive Is Dead."

106 Ibid.

107 Daniel Wincott, "Federalism and the European Union: The Scope and Limits of the Treaty of Maastricht," *International Political Science Review* 17, no. 4 (1996): 403–15.

108 *Amended Proposal for a Council Directive on the Legal Protection of Biotechnology Inventions*, at 4, COM (92) 589 final (December 4, 1992).

109 Group of Advisers on the Ethical Implications of Biotechnology to the European Commission, "Opinion on Ethical Questions Arising from the Commission Proposal for a Council Directive on Legal Protection for Biotechnological Inventions" (September 30, 1993), on file with author; Kamstra, et al., *Patents on Biotechnological Inventions*.

110 *Amended Proposal for a Council Directive on the Legal Protection of Biotechnology Inventions*, at 24, COM (92) 589 final (December 16, 1992).

111 Ibid.

112 Steve Emmott, "No Patents on Life: The Incredible Ten-year Campaign Against the Euro-

pean Patent Directive," in *Redesigning Life? The Worldwide Challenge to Genetic Engineering*, ed. Brian Tokar (New York: Zed Books, 2001).

113 British Society for Human Genetics, *Patenting of Human Gene Sequences and the EU Draft Directive: Statement by the British Society for Human Genetics*, September 1997. Online at http://www.bshg.org.uk/Official%20Docs/patent_eu.htm; Tony Andrews, et al., "As researchers or clinical scientists . . ." Letter to European Parliamentarians, July 14, 1997; Wendy Watson, patient advocate, personal Interview with author, June 1998.

114 Barbara Dinham, Pesticides Trust, et al., to Peter Stevenson, Compassion in World Farming, et al., "Suggested Plans for Meeting with John Battle MP . . . ," July 24, 1997, letter on file with author.

115 Alan Simpson, Nicholas Hildyard, and Sarah Sexton, "No Patents on Life: A Briefing on the Proposed EU Directive on the Legal Protection of Biotechnological Inventions," *Corner House Briefing 01*, September 1, 1997. Online at http://www.thecornerhouse.org.uk/resource/no-patents-life; Church of Scotland, "Church of Scotland Urges European Parliament to Prevent Patenting Human Genes" (press release, May 22, 1997).

116 Nigel Hawkes, "Euro MPs Turn Down Life-Form Patent Law," *The Times*, March 2, 1995; Daniel Green, "Parliament Scuppers a New Patents Directive," *Financial Times*, March 3, 1995, 4; Charles Bremmer, "Euro-MPS Clear Way for Genetic Patents," *The Times*, May 13, 1998; Katherine Butler and Charles Arthur, "Anger as Europe Votes to 'Sell Off' Genes," *The Independent*, May 13, 1998, 1; Emmott, "No Patents on Life."

117 Remarks of Alexander Langer, EUR. PARL. DEB. (4-458) 57 (March 1, 1995).

118 Remarks of Roberto Mezzaroma, EUR. PARL. DEB. (4-458) 39 (March 1, 1995).

119 Ibid.

120 Ibid.

121 Emmott, "No Patents on Life."

122 *Proposal for a European Parliament and Council Directive on the Legal Protection of Biotechnological Inventions*, at 13, COM (95) 661 final (December 13, 1995).

123 Institute of Professional Representatives before the EPO, COM 13830 (September 1996): 2, on file with author.

124 Hilary Bower, "Whose Genes Are They Anyway? Genes Part 1: The Patent Debate," *The Independent*, November, 8 1997, 52.

125 *Report on the Proposal for a European Parliament and Council Directive on the Legal Protection of Biotechnological Inventions*, at 44, COM (95) 661 (June 25, 1997).

126 *Report on the Proposal for a European Parliament and Council Directive on the Legal Protection of Biotechnological Inventions*, at 56, COM (95) 661 (June 25, 1997).

127 See, for example, Remarks of Leen van der Waal, EUR. PARL. DEB. (4-504) 35 (July 15, 1997).

128 Remarks of Ulla Sandbaek, EUR. PARL. DEB. (4-519) 9 (May 11, 1998).

129 Committee on Legal Affairs, "Debates of the European Parliament," European Parliament (July 15, 1997), on file with author.

130 Remarks of Roberto Mezzaroma, EUR. PARL. DEB. (4-504) 36 (July 15, 1997).

131 Remarks of Pat Cox, EUR. PARL. DEB. (4-504) 36 (July 15, 1997).

132 Bernard Steunenberg and Jacques Thomassen, eds., *The European Parliament: Moving toward Democracy in the EU* (New York: Rowman & Littlefield, 2002).

133 Council Directive (EC) No. 98/44 of 6 July 1998, 1998 O.J. (L 213) 13.

134 Thomas Pogge, Matthew Rimmer, and Kim Rubenstein, *Incentives for Global Public Health: Patent Law and Access to Essential Medicines* (New York: Cambridge University Press, 2010).

CHAPTER THREE

1 Karen Rader, *Making Mice: Standardizing Animals for American Biomedical Research, 1900–1955* (Princeton, NJ: Princeton University Press, 2004).

2 In re Allen, 846 F.2d 77, No. 87-1393 (Fed. Cir. March 14, 1988); In re Bergy, 596 F.2d 952, No. 76-712 77-535 (C.C.P.A. March 29, 1979). Online at http://openjurist.org/846/f2d/77/in-re-allen; https://casetext.com/case/in-re-application-of-bergy-2.

3 Donald J. Quigg, "Animals—Patentability," *Official Gazette* 1077 (April 21, 1987): 24.

4 Jeremy Rifkin, "Patenting Forms of Animal Life: Is Nature Just a Form of Private Property?" *New York Times*, April 26, 1987, 2.

5 Keith Schneider, "Senator Asks Halt to Animal Patents," *New York Times*, May, 15, 1987; Keith Schneider, "Agency and Congress Face Clash Over Patenting of Animals," *New York Times*, July, 23, 1987.

6 Daniel J. Kevles, "Of Mice & Money: The Story of the World's First Animal Patent," *Daedalus* 131, no. 2 (2002): 78–88.

7 *Patents and the Constitution: Transgenic Animals: Hearing Before the H. Subcomm. on Courts, Civil Liberties, and the Admin. of Justice of the Comm. on the Judiciary*, 100th Cong. 56 (1987).

8 Ibid., 351.

9 Ibid., 315.

10 Ibid., 485–86.

11 Ibid., 394.

12 Ibid., 102.

13 Ibid., 464.

14 Ibid., 6.

15 Ibid., 373.

16 Ibid., 373.

17 Andrew Abbott, *The System of Professions: An Essay on the Division of Expert Labor* (Chicago: University of Chicago Press, 1988).

18 Ibid., 136.

19 Ibid., 435.

20 Luigi Palombi, "The Patenting of Biological Materials in the United States: A State of Policy Confusion," *Perspectives on Science* 23, no. 1 (2015): 35–65.

21 Thomas F. Gieryn, *Cultural Boundaries of Science: Credibility on the Line* (Chicago: University of Chicago Press, 1999).

22 *Transgenic Animal Patent Reform Act of 1989: Hearing on H.R. 1556 Before the H. Subcomm. on Courts, Intellectual Property, and the Admin. of Justice of the Comm. on the Judiciary*, 101st Cong. 1–685 (1989).

23 Ibid.; Patent Competitiveness and Technological Innovation Act of 1990, H.R. 5598, 101st Cong. §2169, S1612 (1990); Life Patenting Moratorium Act of 1993, S. 387, 103rd Cong. §1792 (1993).

24 Edmund L. Andrews, "Patents; Applications for Animals Up Sharply," *New York Times*, April 22, 1989, 36.

25 Nan Aron, *Liberty and Justice for All: Public Interest Law in the 1980s and Beyond* (Boulder, CO: Westview Press, 1989).

26 Stuart A. Scheingold and Austin Sarat, *Something to Believe In: Politics, Professionalism, and Cause Lawyering* (Stanford, CA: Stanford University Press, 2004).

27 Louis G. Trubek, "Public Interest Law: Facing the Problems of Maturity," *University of Arkansas—Little Rock Law Review* 33 (2011): 417–33.

28 "Victories," Animal Legal Defense Fund. Online at http://aldf.org/article.php?list=type& type=81.

29 The ALDF's archives on the case include multiple references to Rifkin's work, particularly his leadership in the congressional hearings on animal patents (documents on file with author).

30 Animal Legal Defense Fund, "Animal Rights Groups Sue U.S. Patent Office" (press release, San Rafael, CA, July 28, 1988), on file with author.

31 Edmund L. Andrews, "Patents; Lawsuit Challenges Patenting of Animals," *New York Times*, July 30, 1988, 34.

32 Ibid.

33 Joanna Grisinger, *The Unwieldy American State: Administrative Politics since the New Deal* (New York: Cambridge University Press, 2012).

34 Herbert P. Kitschelt, "Political Opportunity Structures and Political Protest: Anti-Nuclear Movements in Four Democracies," *British Journal of Political Science* 16, no. 1 (1986): 57–85.

35 Lujan v. Defenders of Wildlife, 504 U.S. 555 (1992).

36 Animal Legal Defense Fund et al. v. Donald J. Quigg and C. William Verity, 900 F. 2d 195, 6 (9th Cir. 1990) C88 2938, *United States District Court for the Northern District of California* (1988), on file with author.

37 Ibid.

38 Animal Legal Defense Fund et al. v. Donald J. Quigg and C. William Verity, 932 F.2d 920, No. 90-1364, (1991): 40.

39 Ibid.

40 Stuart Newman, "Chimeric Embryos and Animals Containing Human Cells," US Patent Application No. 08/993,564 (filed December 18, 1997).

41 James F. Childress, "Deliberations of the Human Fetal Tissue Transplantation Research Panel," in *Biomedical Politics*, ed. Kathi J. Hanna (Washington, DC: National Academies Press, 1991); Susan E. Kelly, "Public Bioethics and Publics: Consensus, Boundaries, and Participation in Biomedical Science Policy," *Science, Technology, and Human Values* 28, no. 3 (2003): 339–64.

42 Kali Murray, *A Politics of Patent Law: Crafting the Participatory Patent Bargain* (New York: Routledge, 2012).

43 John Mackenzie, "A Closer Look," *ABC World News Tonight* with Peter Jennings, April 2, 1998; Rick Weiss, "Patent Sought on Making of Part-Human Creatures," *Washington Post*, April 2, 1998; Daniel Zwerdling, "Humanimals," *Weekend All Things Considered*, National Public Radio, April 5, 1998.

44 Stuart A. Newman, "My Attempt to Patent a Human-Animal Chimera," *L'Observatoire de la Génétique* 27 (April–May 2006). Online at https://www.nymc.edu/sanewman/PDFs/L %27Observatorie%20Genetique_chimera.pdf.

45 Commissioner Bruce Lehman, "Media Advisory," 1998. (on file with author); USPTO,

"Facts on Patenting Life Forms Having a Relationship to Humans" (press release 98-6, April 1, 1998). Online at http://www.uspto.gov/about-us/news-updates/facts-patenting-life-forms-having-relationship-humans.

46 Mark Dowie, "Gods and Monsters," *Mother Jones* (January/February 2004). Online at http://www.motherjones.com/politics/2004/01/gods-and-monsters.

47 Jenna Greene, "He's Not Just Monkeying Around," *Legal Times* (August 16, 1999): 25.

48 Ibid.

49 Chimeric Embryos and Animals Containing Human Cells, U.S. Patent No. 08/993,564 (filed Dec. 18, 1997).

50 Weiss, "Patent Sought on Making of Part-Human Creatures."

51 Chimeric Embryos and Animals Containing Human Cells, U.S. Patent No. 08/993,564 (filed Dec. 18, 1997); Patrick Coyne to Deborah Crouch, "Amendment and Response," Patent Application No. 08/993,564 (1999).

52 Ibid.; Deborah Crouch to Patrick Coyne, "Final Rejection," Patent Application No. 08/9 93,564 (1999).

53 Ibid.

54 Mark Dowie, "Gods and Monsters," *Mother Jones*, January/February 2004, 48–53.

55 International Center for Technology Assessment, PatentWatch Project to Assistant Commissioner for Patents, "Dear Sir or Madam: Reexamination under 35 U.S.C. §§311–318 and 37 CFR §§1.902-1997 is respectfully requested of U.S. Patent No. 6,444,872 . . . ," (2004), on file with author; American Anti-Vivisection Society, International Center for Technology Assessment, and Alternatives Research and Development Foundation, *In Re U.S. Patent No. US-6,924,413-B2* (2007). Online at http://www.icta.org/files/2012/03/chal lenge_rabbitpatent.pdf.

56 Animal Anti-Vivisection Society, "Stop Animal Patents: Animals Are NOT Inventions," *Website* (October 4, 2011). Online at http://archive.is/ZVMg.

57 Mark D. Janis, "Rethinking Reexamination: Toward a Viable Administrative Revocation System for US Patent Law," *Harvard Journal of Law and Technology* 11, no. 1 (1997): 1–122.

58 International Center for Technology Assessment, PatentWatch Project to Assistant Commissioner for Patents, "Dear Sir or Madam: Reexamination under 35 U.S.C. §§311–318 and 37 CFR §§1.902-1997 is respectfully requested of U.S. Patent No. 6,444,872 . . . ," (2004), on file with author; American Anti-Vivisection Society, International Center for Technology Assessment, and Alternatives Research and Development Foundation, *In Re U.S. Patent No. US-6,924,413-B2* (2007), accessed May 3, 2012, http://www.icta.org/files /2012/03/challenge_rabbitpatent.pdf.

59 International Center for Technology Assessment, PatentWatch Project to Assistant Commissioner for Patents, "Dear Sir or Madam: Reexamination under 35 U.S.C. §§311–318 and 37 CFR §§1.902-1997 is respectfully requested of U.S. Patent No. 6,444,872 . . . ," (2004), on file with author.

60 American Anti-Vivisection Society, International Center for Technology Assessment, and Alternatives Research and Development Foundation, *In Re U.S. Patent No. US-6,924,413-B2* (2007). Online at http://www.icta.org/files/2012/03/challenge_rabbitpatent .pdf, 31.

61 AAVS, International Center for Technology Assessment, Alternatives Research & Development Foundation, Patent on Rabbits Rejected (press release, March 4, 2008). Online at http://www.icta.org/files/2012/03/pressrelease_Patent_on_Rabbits_Rejected.pdf.

62 AAVS and PatentWatch Project, "Patent on Beagle Dogs Canceled" (press release, May 27, 2004), on file with author.

63 In interviews, PTO personnel cited both this case and the public ridicule it endured in 1999 after granting a patent on a frozen peanut butter sandwich (Seth Shulman, "PB&J Patent Punch-Up," *MIT Technology Review*, May 1, 2001), as providing the impetus for the SAWS system.

64 Memorandum from Janice A. Falcone et al., Group Director, United States Patent and Trademark Office, on TC 2800 Guidelines for Sensitive Application Warning System (SAWS) Program Reminder, to the TC 2800 managers (March 27, 2006), on file with author.

65 Interview with PTO Employee A, June 18, 2009.

66 Ibid.

67 Memorandum from Janice A. Falcone et al., Group Director, United States Patent and Trademark Office, on TC 2800 Guidelines for Sensitive Application Warning System (SAWS) Program Reminder, to the TC 2800 managers (March 27, 2006), on file with author.

68 Interview with PTO Employee F, June 12, 2009.

69 Interview with PTO Employee D, April 3, 2009; interview with PTO Employee F, June 12, 2009.

70 Interview with PTO Employee G, July 21, 2009.

71 US Patent and Trademark Office to Shobita Parthasarathy, "Re: Freedom of Information Act (FOIA) Request No. 08-076," 2007.

72 "Sensitive Application Warning System," Web site. Online at http://www.uspto.gov /patent/initiatives/patent-application-initiatives/sensitive-application-warning -system.

73 "Secret PTO Program Subjects Apps to Heightened Scrutiny," Law360.com Web site. Online at http://www.law360.com/articles/600378.

74 "SAWS gets the axe," World Intellectual Property Review Web site. Online at http://www .worldipreview.com/article/saws-gets-the-axe; Alyssa Bereznak, "The U.S. Government Has a Secret System for Stalling Patents." Online at https://www.yahoo.com/tech/the-u -s-government-has-a-secret-system-for-104249688314.html.

75 Fiona Murray, "The Oncomouse That Roared: Hybrid Exchange Strategies as a Source of Distinction at the Boundary of Overlapping Institutions," *American Journal of Sociology* 116, no. 2 (2010): 341–88.

76 European Patent Office, "Communication Pursuant to Article 96(2) and Rule 51(2) EPC. Re: Application 85304490.7-2105," April 20, 1988, 3, on file with author.

77 Ibid.

78 Ibid.

79 European Patent Office, "Decision to Refuse a European Patent Application: Summary of Facts and Submissions. Re: Application 85304490.7-2105," July 14, 1989, 11, on file with author.

80 Letter to the European Parliament, "Einspruch durch Dritte nach Artikel 115 [Opposition by a third party pursuant to Article 115]," April 25, 1990 (Received by the EPO April 30, 1990).

81 Ibid., 21.

82 European Patent Office, "Decision to Grant a European Patent Pursuant to Article 97(2) EPC. Re Application No. 85304490.7-2015," April 3, 1992, on file with author.

83 See, for example, Gen-ethisches Netzwerk Osterreich to Europäisches Patentamt, "Wir widerrufen hiermit frühere Vollmachten in Sachen der obenbezeichneten Anmeldun bzw. Des obenbezeichneten Patents [We hereby revoke all previous authorizations in respect of the above application and the above-identified patent]," Re: Application No: 84302533.9, March 19, 1992; Deutschen Tierschutzbundes e.v. to Europäisches Patentamt, "Einspruchsschrift [Notice of Opposition] Re: Application Number 84302533.9," (December 1989); Greenpeace UK to European Patent Office, "Notice of Opposition Against a European Patent," Patent No. 0242236, July 8, 1990.

84 Dietmar Harhoff and Markus Reitzig, "Determinants of Opposition against EPO Patent Grants—The Case of Biotechnology and Pharmaceuticals," *International Journal of Industrial Organization* 22, no. 4 (2004): 443–80.

85 Plant genetic systems v. Greenpeace Ltd., Patent No. 87400141.5, EPO Technical Board of Appeal decision, T356/93, February, 21, 1995, 9.

86 Stewart and Leder, Method for Producing Transgenic Animals, European Patent Register EP0169672, filed June 24, 1985, and revoked June 13, 2015.

87 Susan Watts, "Backlash Blocks 'Invention' of Animals," *The Independent*, November 30, 1992, 4; Tom Wilkie, "Genetic Patent on Cancer Mouse Faces Opposition," *The Independent* January 13, 1993, 7.

88 See, for example, Compassion in World Farming and British Union for the Abolition of Vivisection, "Oncomouse: Opposition," under Part V of the European Patent Convention, Case T 19/90-3.3.2, Application No. 85 304 490.7, Publication No. 0 169 672, January 8 1993, on file with author.

89 "Ich schliesse mich dem Einspruch gegen Patent 0169672 B1 (Krebsmaus) an [I agree with the objection to patent 0169672 B1 (cancer mouse)]," Received by the EPO, February 1993.

90 The Eurobarometer survey is conducted twice per year on behalf of the European Commission's Directorate General on Audio-Visual, Information, Communication, and Culture. A representative sample of the population in each EU member state is interviewed face-to-face using an identical set of questions; on average, the regular sample is one thousand people in each member state. "Public Opinion Analysis—Homepage—European Commission," European Commission. Online at http://ec.europa.eu/public _opinion/index_en.htm.

91 European Parliament, "Common Motion for a Resolution . . . on First European Patent on Animals," February 10, 1993, PE 170.177/RC1, 170.199/RC1, 170.228/RC1, on file with author.

92 See European Patent 0169672 prosecution, EPO Register Plus database.

93 See, for example, S. Hamberger, H. Jaresch, and C. Then, "Einspruch gegen das Europäische Patent 0169672 [Opposition against the European Patent 0169672]," Kein Patent Auf Leben, February 12, 1993.

94 Ibid.

95 Benno Vogel, "OncoMouse™: Eine Recherche zur medizinischen und kommerziellen Bedeutung der Harvard-Krebsmäuse [A review of the medical and commercial importance of the Harvard Oncomouse]," Greenpeace, November 2, 2001.

96 Ibid.

97 Interview with EPO Employee G, July 22, 2008.

98 Interview with EPO Employee B, August 5, 2008.

99 European Patent Office, Minutes of the Oral Proceedings before the Opposition Division, "Re: Application No. 85304490.7," EPO Register Plus Database, November 21–24 (1995):18, on file with author.

100 Interview with EPO Employee A, June 25, 2008.

101 The President and Fellows of Harvard College v. British Union for the Abolition of Vivisection et al., Patent No. 85304490.7, EPO Technical Board of Appeal decision, T 0315/03, July 6, 2004, 109.

102 Ibid., 72.

103 Ibid., 94.

104 Ibid., 125.

105 Ibid.

106 Bryan Karet, "Patents and Ethics in the Context of Modern Technology," *EPO Gazette* (Munich, Germany: European Patent Office, 1992): 13–17.

107 "Patente auf Lebewesen und Gene [Patents on creatures and genes]," Frankfurter Allgemeine Zeitung, July 14, 1993.

108 Sinasi Bayrak, "Ethical Issues in Engineering and Science," *EPO Gazette* (Munich, Germany: European Patent Office, July–August 2007): 15.

109 Interview with EPO Employee C, July 2, 2008.

110 Interview with EPO Employee F, July 27, 2009.

111 Interview with EPO Employee C, July 2, 2008.

112 Interview with EPO Employee A, June 25, 2008.

113 Ibid.

114 Ibid.

115 Interview with EPO Employee C, July 2, 2008.

CHAPTER FOUR

1 Suzanne Holland, Karen Lebacqz, and Laurie Zoloth, *The Human Embryonic Stem Cell Debate: Science, Ethics, and Public Policy* (Cambridge, MA: MIT Press, 2001); D. C. Wertz, "Embryo and Stem Cell Research in the United States: History and Politics," *Gene Therapy* 9, no. 11 (2002): 674–78; Charis Thompson, *Good Science: The Ethical Choreography of Stem Cell Research* (Cambridge, MA: MIT Press, 2013); Brian Salter, "Bioethics, Politics, and the Moral Economy of Human Embryonic Stem Cell Science: The Case of the European Union's Sixth Framework Programme," *New Genetics and Society* 26, no. 3 (2007): 269–88.

2 John Gearheart, "New Potential for Human Embryonic Stem Cells," *Science* 282, no. 5391 (1998): 1061–62.

3 Herbert Gottweis, "Stem Cell Policies in the United States and in Germany: Between Bioethics and Regulation," *Policy Studies Journal* 30, no. 4 (2002): 444–69; Noëlle Lenoir, "Europe Confronts the Embryonic Stem Cell Research Challenge," *Science* 287, no. 5457 (2000): 1425–27; Samantha Halliday, "A Comparative Approach to the Regulation of Human Embryonic Stem Cell Research in Europe," *Medical Law Review* 12 (2004): 40–69; Sheila Jasanoff, *Designs on Nature: Science and Democracy in Europe and the United States* (Princeton, NJ: Princeton University Press, 2005).

4 Robert Cook-Deegan, *The Gene Wars: Science, Politics, and the Human Genome* (New York: W. W. Norton, 1994).

5 Mark D. Adams, Jenny M. Kelley, Jeannine D. Gocayne, Mark Dubnick, Mihael H. Polumeropoulos, Hong Xiao, Carl R. Merril, et al., "Complementary DNA Sequencing: Expressed Sequence Tags and Human Genome Project," *Science* 252 (1991): 1651–56.

6 Charles R. Cantor and Cassandra L. Smith, *Genomics: The Science and Technology Behind the Human Genome Project* (Hoboken, NJ: John Wiley & Sons, 2004).

7 Leslie Roberts, "Genome Patent Fight Erupts," *Science* 254, no. 5029 (1991): 184–86.

8 *The Genome Project: The Ethical Issues of Gene Patenting: Hearing Before the S. Subcomm. on Patents, Copyrights and Trademarks of the Comm. on the Judiciary United States*, 102nd Cong. 52–53 (1992).

9 Rebecca S. Eisenberg, "Patenting the Human Genome," *Emory Law Journal* 39 (1990): 721–45.

10 Ibid.

11 Leslie Roberts, "NIH Gene Patents, Round Two," *Science* 255, no. 5047 (1992): 912–13; Roberts, "Genome Patent Fight Erupts"; Mark Blaxter, "Patenting of Genes," *Nature* 355, no. 6356 (1992): 104; Thoru Pederson, "NIH Patent Rights," *Nature* 358, no. 6288 (1992): 617; Edmund L. Andrews, "US Seeks Patent on Genetic Codes, Setting Off Furor," *New York Times*, October 21, 1991. Online at http://www.nytimes.com/1991/10/21/us/us-seeks-patent-on-genetic-codes-setting-off-furor.html.

12 "Fresh Debate Over the Life-form Ruling," *Chemical Week*, August 6, 1980, 47; Paul S. Naik, "Biotechnology Through the Eyes of An Opponent: The Resistance of Activist Jeremy Rifkin," *Virginia Journal of Law and Technology* 5 (2000): 5–15.

13 Roberts, "Genome Patent Fight Erupts."

14 The American Society of Human Genetics, "Position Paper on Patenting of Expressed Sequence Tags," November, 1991. Online at http://www.ashg.org/pdf/policy/ASHG_PS_November1991.pdf.

15 Thomas F. Gieryn, *Cultural Boundaries of Science: Credibility on the Line* (Chicago: University of Chicago Press, 1999); Andrew Abbott, *The System of Professions: An Essay on the Division of Expert Labor* (Chicago: University of Chicago Press, 1988).

16 Sheila Jasanoff and Sang-Hyun Kim, "Containing the Atom: Sociotechnical Imaginaries and Nuclear Power in the United States and South Korea," *Minerva* 47, no. 2 (2009): 119–46.

17 *The Genome Project: The Ethical Issues of Gene Patenting: Hearing Before the S. Subcomm. on Patents, Copyrights and Trademarks of the Comm. on the Judiciary United States*, 102nd Cong. 26 (1992).

18 Ibid., 164.

19 Ibid., 7.

20 Ibid., 1–244.

21 Alfred E. Kahn, "Fundamental Deficiencies of the American Patent Law," *American Economic Review* 30, no. 3 (1940): 475–91; Nathan Machin, "Prospective Utility: A New Interpretation of the Utility Requirement of Section 101 of the Patent Act," *California Law Review* 87, no. 2 (1999): 4231–456.

22 United States Patent and Trademark Office, Department of Commerce, "Utility Examination Guidelines," *Federal Register* 66, no. 4 (2001): 1092–99.

23 Ibid., 1094.

24 Ibid.

25 James E. Bessen, "Patent Thickets: Strategic Patenting of Complex Technologies." Online at http://ssrn.com/abstract=327760; Elon Musk, "All Our Patent Are Belong to You," Online at http://www.teslamotors.com/blog/all-our-patent-are-belong-you.

26 David Blumenthal, "Growing Pains for New Academic/Industry Relationships," *Health Affairs* 13, no. 3 (1994): 176–93; Keith Aoki, "Authors, Inventors and Trademark Owners: Private Intellectual Property and the Public Domain," *Columbia-VLA Journal of Law and the Arts* 18, nos. 1 and 2 (1992): 191–267; James Boyle, "A Theory of Law and Information: Copyright, Spleens, Blackmail, and Insider Trading," *California Law Review* 80, no. 6 (1992): 1413–540.

27 Rebecca Eisenberg, "Patents and the Progress of Science: Exclusive Rights and Experimental Use," *University of Chicago Law Review* 56, no. 3 (1989): 1017–86; Rebecca Eisenberg, "Proprietary Rights and the Norms of Science in Biotechnology Research," *Yale Law Journal* 97, no. 2 (1987): 177–231.

28 Whittemore v. Cutter, 29 F. Cas. 1120 (U.S. App. 1813).

29 Eisenberg, "Patents and the Progress of Science."

30 Michael A. Heller and Rebecca S. Eisenberg, "Can Patents Deter Innovation? The Anticommons in Biomedical Research," *Science* 280, no. 5364 (1998): 698–701.

31 See, for example, Arti Rai, "Regulating Scientific Research: Intellectual Property Rights and the Norms of Science," *Northwestern University Law Review* 94, no. 1 (1999): 77–152; Robert P. Merges, "Property Rights Theory and the Commons: The Case of Scientific Research," *Social Philosophy and Policy* 13, no. 2 (1996): 145–67.

32 See, for example, Jon F. Merz, "Disease Gene Patents: Overcoming Unethical Constraints on Clinical Laboratory Medicine," *Clinical Chemistry* 45, no. 3 (1999): 324–30; Shobita Parthasarathy, *Building Genetic Medicine: Breast Cancer, Technology, and the Comparative Politics of Health Care* (Cambridge, MA: MIT Press, 2007); Alessandra Colaianni, Subhashini Chandrasekharan, and Robert Cook-Deegan, "Impact of Gene Patents and Licensing Practices on Access to Genetic Testing and Carrier Screening for Tay-Sachs and Canavan Disease," *Genetics in Medicine* 23, no. 1s (2010): S5-S14; Jon F. Merz, Antigone G. Kriss, Debra G. B. Leonard, and Mildred K. Cho, "Diagnostic Testing Fails the Test," *Nature* 415, no. 6872 (2002): 577–79.

33 See, for example, Anna Schissel, Jon F. Merz, and Mildred K. Cho, "Survey Confirms Fears about Licensing of Genetic Tests," *Nature* 402, no. 6758 (1999): 118; Mildred K. Cho, Samantha Illangasekare, Meredith A. Weaver, Debra G. B. Leonard, and Jon F. Merz, "Effects of Patents and Licenses on the Provision of Clinical Genetic Testing Services," *Journal of Molecular Diagnostics* 5, no. 1 (2003): 3–8.

34 See, for example, Heidi Williams, "Intellectual Property Rights and Innovation: Evidence from the Human Genome," *Journal of Political Economy* 121, no. 1 (2013): 1–27; John P. Walsh, Ashish Arora, and Wesley M. Cohen, "Working Through the Patent Problem," *Science* 299 (2003): 1021; John P. Walsh, Ashish Arora, and Wesley Cohen, "Effects of Research Tool Patents and Licensing on Biomedical Innovation," in *Patents in the Knowledge-Based Economy* (Washington, DC: National Academies Press, 2003); Christopher M. Holman, "The Impact of Human Gene Patents on Innovation and Access: A Survey of Human Gene Patent Litigation," *UMKC Law Review* 76, no. 2 (2007–2008): 295–361; David C. Mowery, Richard R. Nelson, Bhaven N. Sampat, and Arvids Ziedonis, "The Growth of Patenting and Licensing by US Universities: An Assessment of the Effects of

the Bayh-Dole Act of 1980," *Research Policy* 20, no. 1 (2001): 99–119; Fiona Murray and Scott Stern, "Do Formal Intellectual Property Rights Hinder the Free Flow of Scientific Knowledge? An Empirical Test of the Anti-Commons Hypothesis," *Journal of Economic Behavior & Organization* 63, no. 4 (2007): 648–87.

35 Stephen Hilgartner, *Science on Stage: Expert Advice as Public Drama* (Stanford, CA: Stanford University Press, 2000).

36 Wesley M. Cohen and Stephen A. Merrill, eds. *Patents in the Knowledge-Based Economy* (Washington, DC: National Academies Press, 2003).

37 Sandra Schmieder, "Scope of Biotechnology Inventions in the United States and in Europe—Compulsory Licensing, Experimental Use and Arbitration: A Study of Patentability of DNA-Related Inventions with Special Emphasis on the Establishment of an Arbitration Based Compulsory Licensing System," *Santa Clara High Technology Law Journal* 21, no. 1 (2004): 164–234; Kevin Iles, "A Comparative Analysis of the Impact of Experimental Use Exemptions in Patent Law on Incentives to Innovate," *Northwestern Journal of Technology and Intellectual Property* 4, no. 1 (2005): 61–82.

38 Trevor Cook, "A European Perspective as to the Extent to Which Experimental Use, and Certain Other Defences to Patent Infringement, Apply to Differing Types of Research." Online at http://www.ipeg.com/_UPLOAD%20BLOG/Experimental%20Use%20for%20IPI%20Chapters%201%20to%209%20Final.pdf.

39 Peter Ruess, "Accepting Exceptions? A Comparative Approach to Experimental Use in U.S. and German Patent Law," *Marquette Intellectual Property Law Review* 10, no. 1 (2006): 82–110.

40 Janice M. Mueller, "No 'Dilettante Affair': Rethinking the Experimental Use Exception to Patent Infringement for Biomedical Research Tools," *Washington Law Review* 76, no. 1 (2001): 1–66.

41 Rick Weiss, "A Crucial Human Cell Is Isolated, Multiplied: Embryonic Building Block's Therapeutic Potential Stirs Debate," *Washington Post*, November 6, 1998.

42 Fiona Murray, "The Stem-Cell Market—Patents and the Pursuit of Scientific Progress," *New England Journal of Medicine* 356, no. 23 (2007): 2341–43.

43 *Stem Cell Research: Hearing Before the S. Subcomm. of the Comm. on Appropriations*, 105th Cong. 2 (1998).

44 Ibid., 114–15.

45 Ibid., 85.

46 Ibid., 81.

47 National Institutes of Health, "Human Embryonic Stem Cell Policy under Former President Bush (Aug. 9, 2001–Mar. 9, 2009)." Online at http://stemcells.nih.gov/policy/pages/2001policy.aspx.

48 Judith A. Johnson and Erin D. Williams, "Stem Cell Research: State Initiatives," Congressional Research Service Reports RL33524, last modified January 1, 2006. Online at http://digitalcommons.unl.edu/crsdocs/38/.

49 Mark B. Brown and David H. Guston, "Science, Democracy, and the Right to Research," *Science and Engineering Ethics* 15, no. 3 (2009): 1; Atossa M. Alavi, "Stem Cell Compromise: A Wolf in Sheep's Clothing, Constitutional Implications of the Bush Plan," *Health Matrix* 13, no. 1 (2003): 181.

50 David B. Resnik, "Privatized Biomedical Research, Public Fears, and the Hazards of Government Regulation: Lessons From Stem Cell Research," *Health Care Analysis* 7, no. 3

(1999): 273–87; J. M. Golden, "WARF's Stem Cell Patents and Tensions between Public and Private Sector Approaches to Research," *Journal of Law, Medicine and Ethics* 38, no. 2 (2010): 314–31; Brown and Guston, "Science, Democracy, and the Right to Research."

51 *Stem Cells, 2001: Hearing before the S. Subcomm. of the Comm. on Appropriations*, 107th Cong. 100 (2001).

52 Ibid., 88.

53 Ibid., 123, 150.

54 CONG. REC. S5515 (June 13, 2002).

55 Biotechnology Industry Organization, "New Patent Legislation Sets Dangerous Precedent and Stifles Research," Fact Sheet, September 2, 2003.

56 CONG. REC. H2274 (March 24, 2003).

57 Patent Watch, "Executive Summary of Patent Watch's Discovery of the First US Patent for Reproductive Human Cloning" (press release), May 16 2002; Patent Watch, "Fetal Dependent Patents: An Interim Report," October 5 2005.

58 Roger G. Noll, "The Politics and Economics of Implementing State-Sponsored Embryonic Stem-Cell Research," SIEPR Discussion Paper No. 04-28, June 2005. Online at http://www.researchgate.net/profile/Roger_Noll/publication/241767339_The_Politics_and_Economics_of_Implementing_State-Sponsored_Embryonic_Stem-Cell_Research/links/0f31753348f5b2b8bf000000.pdf.

59 Douglas Johnson, Legislative Director, National Right to Life Committee, to Senators, "The National Right to Life Committee (NRLC) urges you to support the pending Brownback-Ensign amendment to S. 2600," letter of June 17, 2002. Online at http://www.nrlc.org/archive/Killing_Embryos/Senate%20patenting.pdf; Richard M. Doerflinger, "Congressional Impasse on Human Cloning," *National Catholic Bioethics Quarterly* (Spring 2004). Online at http://www.ncbcenter.org/page.aspx?pid=941; National Committee for a Human Life Amendment, "Patenting Human Organisms Ban: Fiscal Year 2005 Commerce/Justice/State Appropriations," modified 2004. Online at http://nchla.org/legissectiondisplay.asp?ID=475.

60 Noll, "Politics and Economics of Implementing State-Sponsored Embryonic Stem-Cell Research."

61 California Research and Cures Initiative (Proposition 71), Cal. Const. article XXXV adding the California Stem Cell Research and Cures/Bond Act to the Health and Safety Code, §125290.10–125290.70 (2004).

62 Ibid.

63 Ibid.

64 See, for example, *Government Patent Policy: Part 1: Hearing on S. 48 Before the S. Subcomm. on Patents, Trademarks, and Copyrights of the Comm. on the Judiciary*, 89th Cong. 327 (1965).

65 Madey v. Duke University, 307 F.3d, 1351 (2002).

66 Intellectual Property Task Force, California Institute of Regenerative Medicine, "Regular Meeting," Transcript, October 25, 2005, 47, on file with author.

67 Ibid., 91.

68 Intellectual Property Study Group, California Council on Science and Technology, "Policy Framework for Intellectual Property Derived from Stem Cell Research in California: Interim Report to the California Legislature, Governor of the State of California,

California Institute for Regenerative Medicine," August 2005. Online at http://www.ccst .us/publications/2005/IPinterim.pdf.

69 Intellectual Property Task Force, California Institute of Regenerative Medicine, "Regular Meeting," Transcript, October 25, 2005, 23, 26, on file with author.

70 Ibid., 130.

71 Barbara M. McGarey and Annette Levey, "Patents, Products, and Public Health: An Analysis of the CellPro March-in Petition," *Berkeley Technology Law Journal* 14, no. 3 (1999): 1095–116; Rebecca S. Eisenberg, "Public Research and Private Development: Patents and Technology Transfer in Government-Sponsored Research," *Virginia Law Review* 82, no. 8 (1996): 1662–727.

72 Intellectual Property Task Force, California Institute of Regenerative Medicine, "Regular Meeting," Transcript, November 22, 2005, on file with author.

73 Ibid.

74 Intellectual Property Task Force, Transcript, October 25, 2005, 150, on file with author.

75 Ibid., 85–87.

76 "Egg Extraction for Stem Cell Research: Issues for Women's Health," Center for Genetics and Society Website, last modified May 25, 2006. Online at http://www.geneticsandsociety .org/article.php?id=950.

77 Biotechnology Industry Organization, "Comments of the Biotechnology Industry Organization on California Institute for Regenerative Medicine's Notice of Proposed Regulation Adoption, California Code of Regulations. Title 17.—Public Health Proposed Regulations: Intellectual Property Policy for Non-Profit Organizations," June 16, 2006, on file with author.

78 David L. Gollaher, President and CEO, California Healthcare Institute, to C. Scott Tocher, Interim Counsel, California Institute for Regenerative Medicine, "Comments to Proposed CIRM Regulation Entitled: Intellectual Property Policy for Non-Profit Organizations . . ." Letter, June 15, 2006, 3, on file with author.

79 Intellectual Property Task Force, California Institute of Regenerative Medicine, "Regular Meeting," Transcript, July 14, 2006, 47, on file with author.

80 Intellectual Property Task Force, California Institute of Regenerative Medicine, "Regular Meeting," Transcript, August 29, 2006, 16, on file with author.

81 "Chapter 6. Intellectual Property and Revenue Sharing Requirements for Non-Profit and For-Profit Grantees," California Institute for Regenerative Medicine Web site. Online at http://www.cirm.ca.gov/our-funding/chapter-6-intellectual-property-and-revenue -sharing-requirements-non-profit-and-profit.

82 Jeanne Loring, "A Patent Challenge for Human Embryonic Stem Cell Research," *Nature Reports Stem Cells*, November 8, 2007. Online at http://www.nature.com/stemcells/2007 /0711/071108/full/stemcells.2007.113.html.

83 Public Patent Foundation (2003–2011), Home Page. Online at http://www.pubpat.org.

84 Crispin Littlehales, "Dan Ravicher," *Nature Biotechnology* 26, no. 4 (2008): 369.

85 See, for example, Public Patent Foundation, "*Knorr-Bremse v. Dana Corporation: United States Court of Appeals for the Federal Circuit, Amicus Brief. 01-1357, -1376, 02-1221, -1256*," on file with author.

86 Public Patent Foundation, "Patent No. 7,029,913," *Attachment to Form PTO-1465: Request for Inter Partes Reexamination* (April 18, 2006): 2, on file with author.

87 "Groups Challenge Stem Cell Patents That Loot Taxpayer Funds and Force Research Overseas," Consumer Watchdog Web site. Online at http://www.consumerwatchdog .org/newsrelease/groups-challenge-stem-cell-patents-loot-taxpayer-funds-and-force -research-overseas.

88 See, for example, US Patent and Trademark Office, "Detailed Action: Reexamination: Granting of Request," Patent No. 5843780, Reexamination Control No 90/008102, September 29, 2006, 4, on file with author.

89 Public Patent Foundation, PTO REJECTS HUMAN STEM CELL PATENTS AT BEHEST OF CONSUMER GROUPS: Re-examination Was Initiated by Foundation for Taxpayer and Consumer Rights and Public Patent Foundation (press release). Online at http://www .pubpat.org/warfstemcellpatentsrejected.htm.

90 Barbara Prainsack and Robert Gmeiner, "Clean Soil and Common Ground: The Biopolitics of Human Embryonic Stem Cell Research in Austria," *Science as Culture* 17, no. 4 (2008): 377–95.

91 Herbert Gottweis and Barbara Prainsack, "Emotion in Political Discourse: Contrasting Approaches to Stem Cell Governance in the USA, UK, Israel, and Germany," *Regenerative Medicine* 1, no. 6 (2006): 823–29.

92 Christoph Then, "Personal Communication," e-mail message to author, July 7, 2008.

93 European Patent Office, "'Öl-Mais'-Patent nach Anhörung im Einspruchsverfahren vor dem Europäischen Patentamt widerrufen/Ergebnis der mündlichen Verhandlung zu einem Patent der Firma DuPont [After opposition hearings, the European Patent Office revokes DuPont's corn oil patent]" (press release, 2003); European Patent Office, "'Neem Tree Oil' Case: European Patent No. 0436257 Revoked" (press release, 2000), on file with author.

94 Interview with European civil society activist, "Personal Communication," e-mail message to author, June 11, 2008.

95 Interview with KPAL staff #2, July 7, 2008.

96 Quirin Schiermeier, "Germany Challenges Human Stem Cell Patent Awarded 'By Mistake,'" *Nature* 404 (2000): 4.

97 Michael Hagmann, "Protest Leads Europeans to Confess Patent Error," *Science* 287, no. 5458 (2000): 1567–69.

98 Jim Bohlen, *Making Waves: The Origin and Future of Greenpeace* (Montreal, Quebec: Black Rose Books, 2000).

99 Rachel Schurman and William A. Munro, *Fighting for the Future of Food: Activists versus Agribusiness in the Struggle Over Biotechnology* (Minneapolis: University of Minnesota Press, 2010); Herbert Gottweis, Brian Salter, and Catherine Waldby, *The Global Politics of Human Embryonic Stem Cell Science* (New York: Palgrave Macmillan, 2009).

100 Interview with KPAL staff #1, June 11, 2008.

101 Deborah Smith, "Approving GM People Was a 'Patent Mistake,'" *Sydney Morning Herald*, February 23, 2000, 3; "Human Cloning Patent Approved 'By Mistake,'" *Vancouver Sun*, February 23, 2000, A14.

102 Schiermeier, "Germany Challenges Human Stem Cell Patent," 3–4.

103 Interview with EPO Official, June 25, 2008.

104 Interview with KPAL staff #2, July 7, 2008; European Patent Office, "EP069351— Isolation, Selection, and Propagation of Animal Transgenic Stem Cells Other Than Em-

bryonic Stem Cells," *European Patent Register*. Online at https://register.epo.org/applica tion?number=EP94913174.

105 Interview with Christoph Then, July 7, 2008.

106 Interview with US patent lawyer, October 31, 2007.

107 Interview with KPAL staff #2, July 7, 2008.

108 "Daniel Alexander QC," 8 New Square: Intellectual Property Web site. Online at http://www.8newsquare.co.uk/members-of-chambers/Daniel+Alexander.

109 Schiermeier, "Germany Challenges Human Stem Cell Patent."

110 Ökumenischer Rat der Kirchen in Österreich, "Einspruch gegen das europäische patent EP 695 351—manipulation der keimbahn [Opposition against European patent EP695351—manipulation of the germ-line]," 1999. EPO Register Plus database, on file with author.

111 Kein Patent Auf Leben, "Sehr Geehrter Herr Wolinski, Beiliegend Möchte Ich Nochmal Acht Dokumente in Das Beschwerdeverfahren Einbringen [Dear Mr. Wolinski, enclosed I would like to introduce 8 more documents to the complaint]," November 4, 2004. EPO Register Plus database, on file with author.

112 Danish Council of Ethics, *Patenting Human Genes and Stem Cells: A Report* (Copenhagen, Denmark: Danish Council of Ethics, 2004).

113 Greenpeace, "Liste Der Einsprechenden Gegen Patent EP 695 351/Sammeleinspruch Greenpeace Stand 17.4.00[List of opponents against EP695351/Greenpeace's collective opposition]," April 18, 2000. EPO Register Plus database, on file with author.

114 See, for example, Bund Deutscher Hebammen e.V., "Einspruch gegen die erteilung des patents (aktenzeichen EP 695351) für ein verfahren zur genetischen manipulation menschlicher embryonen [Opposition to the grant of patent EP695351 for a process of genetically manipulating human embryos]," 2000. EPO Register Plus database, on file with author.

115 Citizen Enquiries Unit, e-mail message to Kathryn Reeves, author's research assistant, December 22, 2014.

116 *European Parliament Resolution on the Decision by the European Patent Office with Regard to Patent No. EP 695 351 Granted on 8 December 1999*, March 30, 2000, O.J. (C378) 95–97.

117 "Universal Declaration on the Human Genome and Human Rights," United Nations Educational, Scientific and Cultural Organization Web site. Online at http://www.unesco.org /new/en/social-and-human-sciences/themes/bioethics/human-genome-and-human -rights/.

118 *European Parliament Resolution on the Decision by the European Patent Office with Regard to Patent No EP 695 351 Granted on 8 December 1999*, 30 March 2000, O.J. (C378) 95–97.

119 George W. Schlich, Mathys & Squire, to European Patent Office, "European Patent No. 0695351 (Granted on Application No. 94913174.2-2105) Patent Owner: University of Edinburgh, Observations in Reply to Opposition," EPO Register Plus database, October 12, 2001, 22, on file with author.

120 Ibid.

121 Ibid., 4.

122 Ibid., 6.

123 European Patent Office, "Minutes of the Oral Proceedings Before the Opposition Division," Application No. 94913174.2, EPO Register Plus database, Patent No. EP0695351, July 22, 2002.

124 European Patent Office, "'Edinburgh' Patent Limited After European Patent Office Opposition Hearing" (press release, July 24, 2002), EPO Register Plus database, on file with the author.

125 *Directive 98/44/EC of the European Parliament and of the Council of 6 July 1998 on the Legal Protection of Biotechnological Inventions*, 30 July 1998, O.J. (L213) 13–21.

126 European Patent Office, "Opposition Hearing on Genetic Stem-Cell Patent at the European Patent Office" (press release, July 18, 2002), on file with author.

127 European Patent Office, "Interlocutory Decision in Opposition Proceedings (Articles 102(3) and 106(3) EPC), Patent No. 0695351," EPO Register Plus database, Application No. 94913174.2-2105, July 21, 2003, on file with author.

128 Ibid.

129 European Patent Office, "Communication Pursuant to Article 96(2) EPC, Application No. 96903521.1-2401, Applicant: Wisconsin Alumni Research Foundation," on file with author.

130 European Patent Office, "Interlocutory Decision of the Technical Board of Appeal 3.3.08, Case No. T 1374/04-3.3.08," November 18, 2005, on file with author.

131 Solidarische Kirche im Rheinland der Leitungskreis to the European Patent Office, "Im Europäischen Patentgesetz warden laut Art 53a Erfindungen vom Patenschutz ausgeschlossen [According to Article 53a in European patent law, patent protection is not allowed] . . ." Amicus Brief for case G2/06, 2006, on file with author.

132 Achille Vernizzi, Cattedra di Statistica Economica to the European Patent Office, "I wish to submit an amicus curiae brief referring to case G2/06," Application No. 96903521.1, September 6, 2006, EPO Register Plus database, on file with author.

133 Roger Kiska, European Centre for Law and Justice, to Registry of the Enlarged Board, European Patent Office, "The European Centre for Law and Justice hereby wishes to submit this *amicus curiae* brief in the case before this enlarged board, G2/06 . . ." September 10, 2006, EPO Register Plus database, on file with author.

134 Pompidou, Alain. G2/06 Comments by the President of the European Patent Office. EP 96903521.1-2401/0770125, EPO Register Plus database, on file with author.

135 Diamond v. Chakrabarty 447 US 303 (1980).

136 epi: Institute of Professional Representatives before the European Patent Office (2006), Amicus curiae brief in the case of G2/06.

137 UK BioIndustry Association, "Amicus Brief on behalf of the UK BioIndustry Association in connection with Referral to the Enlarged Board of Appeal G2/06 following Interlocutory Decision T1374/04 on European Patent Application No 96903521.1," October 31, 2006. EPO Register Plus database, on file with author.

138 Dorothy Nelkin, *Controversy: Politics of Technical Decisions* (Thousand Oaks, CA: Sage, 1992); Daniel Sarewitz, "How Science Makes Environmental Controversies Worse," *Environmental Science and Policy* 7, no. 5 (2004): 385–403.

139 Interview with EPO employee A, June 25, 2008.

140 European Patent Office, "Patents and Clean Energy: Bridging The Gap between Evidence and Policy," Report Funded by the UN Development Programme, EPO, and International Centre for Trade and Sustainable Development. Online at http://www.unep.ch /etb/events/UNEP%20EPO%20ICTSD%20Event%2030%20Sept%202010%20Brussels /Brochure_EN_ganz.pdf; European Patent Office, "Patents and Clean Energy Technologies in Africa," Report funded by the UN Development Programme and EPO, 2013. On-

line at http://www.unep.org/environmentalgovernance/Portals/8/publications/Patents
_Clean_Energy_Technologies_Africa.pdf.

141 See, for example, European Patent Office, "The Melon Patent Case — FAQ," (press release, February 14, 2012), on file with author; European Patent Office, " 'XY/Frozen Sperm' Patent Maintained in Amended Form" (press release, January 30, 2008), on file with author; European Patent Office, "Public Oral Proceedings on the 'Breast Cancer Gene 2' Patent at the European Patent Office" (press release, June 28, 2005), on file with author.

142 Interview with EPO Employee G, July 22, 2008.

143 Ned Stafford, "Oliver Brüstle: Fighting for Stem Cell Research in Germany," *Lancet* 366 (2005): 1521.

144 Alison Abbott, "Fresh Hope for German Stem-Cell Patent Case," *Nature* 462, no. 7271 (2009): 265.

145 Axel Rowohlt, "Patent Protest Erupts Over Human Stem Cell Research," *Deutsche Welle*, last modified December 28, 2006. Online at http://www.dw.de/dw/article/0,,2292175 _page_0,00.html.

146 In 2001, the European Court of Justice had ruled that the BPD was legal, after a challenge from the Dutch, Italian, and Norwegian governments. Kingdom of the Netherlands et al. v. European Parliament and Council, October 9 2001, ECJ C-377/98.

147 Oliver Brüstle v. Greenpeace, October 18, 2011, ECJ C-34/10.

148 A few years later, the ECJ would open a window to the patentability of products and processes related to hESCs, ruling that hESCs could be patented so long as they lacked the capacity to become a human being. But even as it did so, it made its decision on the basis of whether the inventions would affect morality, human dignity, and life, as it saw very different relationships among these things than its US counterparts saw. International Stem Cell Corporation v. Comptroller General of Patents, Designs, and Trademarks, December 18, 2014, ECJ C-364/13.

149 Steve Connor, "Medicine Thrown into Crisis by Stem Cell Ruling; EU Ban on Stem Cell Patents 'Stifles Medical Research,'" *The Independent*, October 19, 2011. Online at http://www.independent.co.uk/news/science/medicine-thrown-into-crisis-by-stem-cell -ruling-2372562.html; Nick Collins, "EU Court Blow to Stem Cell Research," *Daily Telegraph*, October 19, 2011, 18.

CHAPTER FIVE

1 Ronald Labonte, "Global Right to Health Campaign Launched," *British Medical Journal* 331, no. 7511 (2005): 252; Amy Kapczynski, "The Access to Knowledge Mobilization and the New Politics of Intellectual Property" *Yale Law Journal* 117, no. 804 (2008). Online at http://dx.doi.org/10.2139/ssrn.1323525.

2 Manjari Mahajan, "The Right to Health as the Right to Treatment: Shifting Conceptions of Public Health," *Social Research: An International Quarterly* 79, no. 4 (2012): 819–36.

3 Carlos Correa, *Integrating Public Health Concerns into Patent Legislation in Developing Countries* (Geneva, Switzerland: South Centre Press, 2000); Amir Attaran, "How Do Patents and Economic Policies Affect Access to Essential Medicines in Developing Countries?," *Health Affairs* 23, no. 3 (2004): 155–66.

4 Larry B. Stammer and Robert Lee Hotz, "Religious Leaders Oppose Genetic Patents," *Philadelphia Inquirer*, May 18, 1995; Edmund L. Andrews, "Religious Leaders Prepare to Fight Patents on Genes," *New York Times*, May 13, 1995.

5 Robert Dalpé, Louise Bouchard, Anne-Julie Houle, and Louis Bédard, "Watching the Race to Find the Breast Cancer Genes," *Science, Technology, and Human Values* 28, no. 2 (2003): 187–216; Tamar Lewin, "Move to Patent Cancer Gene Is Called Obstacle to Research," *New York Times*, May 21, 1996.

6 "BRCA1 and BRCA2: Cancer Risk and Genetic Testing," National Cancer Institute Web site. Online at http://www.cancer.gov/cancertopics/factsheet/Risk/BRCA.

7 Kevin Davies and Michael White, *Breakthrough: The Race to Find the Breast Cancer Gene* (New York: John Wiley & Sons, 1996).

8 Lewin, "Move to Patent Cancer Gene."

9 Andrea Knox, "The Great Gene Grab: Firms Toss Researchers for a Loop," *Philadelphia Inquirer*, February 13, 2000; Arthur Allen, "Who Owns My Disease?" *Mother Jones*. Online at http://www.motherjones.com/politics/2001/11/who-owns-my-disease.

10 Shobita Parthasarathy, *Building Genetic Medicine: Breast Cancer, Technology, and the Comparative Politics of Health Care* (Cambridge, MA: MIT Press, 2007).

11 Myriad Genetics, Inc., "Myriad Genetics Obtains OncorMed's BRCA1/BRCA2 Genetic Testing Program in Patent Settlement 1998," on file with the author.

12 Myriad's test became available in 1997, more than a decade before the Genetic Information Non-Discrimination Act of 2008 (Public Law 110-233, 122 Stat. 881, enacted May 21, 2008); Soo-Chin Lee, Barbara Bernhardt, and Kathy Helzlsouer, "Utilization of BRCA1/2 Genetic Testing in the Clinical Setting," *Cancer* 94, no. 6 (2002): 1876–85; Chaliki Hemasree, Starlene Loader, Jeffrey Levenkron, Wende Logan-Young, W. Jackson Hall, and Peter Rowley, "Women's Receptivity to Testing for a Genetic Susceptibility to Breast Cancer," *American Journal of Public Health* 85 (1995): 1133–35.

13 Robert Cook-Deegan, Subhashini Chandrasekharan, and Misha Angrist, "The Dangers of Diagnostic Monopolies," *Nature* 458, no. 7237 (2009): 405–6; Julia Carbone, E. Richard Gold, Bhaven Sampat, Subhashini Chandrasekharan, Lori Knowles, Misha Angrist, and Robert Cook-Deegan, "DNA Patents and Diagnostics: Not a Pretty Picture," *Nature Biotechnology* 28 (2010): 784–91.

14 Parthasarathy, *Building Genetic Medicine.*

15 Interview with US Geneticist #4, March 21, 2000; interview with US Geneticist #2, January 11, 2000.

16 National Breast Cancer Coalition, "Gene Patenting: Yes or No?" *Call to Action!* 4, nos. 3 and 4 (1997): 11; David Magnus, Arthur L. Caplan, and Glenn McGee, eds., *Who Owns Life?* (New York: Prometheus Books, 2002); Sharon Terry, "Why Banning Patents Would Hurt Patients," *GeneWatch* 23, no. 5 (2010): 24–25.

17 See, for example, Lori Andrews, "The Gene Patent Dilemma: Balancing Commercial Incentives with Health Needs," *Houston Journal of Health Law and Policy* 2, no. 1 (2002): 65–106; Timothy Caulfield and E. Richard Gold, "Genetic Testing, Ethical Concerns, and the Role of Patent Law," *Clinical Genetics* 57, no. 5 (2001): 370–75; Magnus, Caplan, and McGee, eds., *Who Owns Life?*

18 The NIH's Task Force on Genetic Testing operated from 1995 to 1997. The HHS Secretary's Advisory Committee on Genetic Testing worked from 1998 to 2002. The HHS Secretary's Advisory Committee on Genetics, Health, and Society operated from 2002 to 2010.

19 Secretary's Advisory Committee on Genetic Testing, *Highlights of the Seventh Meeting of the Secretary's Advisory Committee on Genetic Testing November 2–3, 2000.* Online at http://osp.od.nih.gov/sites/default/files/novsacgthigh.pdf.

20 Secretary's Advisory Committee on Genetics, Health, and Society, *Gene Patents and Licensing Practices and Their Impact on Patient Access to Genetic Tests, Report 2010*. Online at http://osp.od.nih.gov/sites/default/files/SACGHS_patents_report_2010.pdf, 4.

21 Wayne W. Grody, "Regulatory Issues in Molecular Diagnostics," in *Molecular Diagnostics: Techniques and Applications for the Clinical Laboratory*, ed. Wayne W. Grody, Robert M. Nakamura, Frederick L. Kiechle, and Charles Strom (Burlington, MA: Academic Press, 2009).

22 Committee on Intellectual Property Rights in Genomic and Protein Research and Innovation, National Research Council, *Reaping the Benefits of Genomic and Proteomic Research: Intellectual Property Rights, Innovation, and Public Health* (Washington, DC: National Academies Press, 2006).

23 Steven Teutsch, SACGHS Chair, to Kathleen Sebelius, HHS Secretary, "In keeping with our mandate to provide advice. . . ." letter, March 31, 2010. Online at https://repository.library.georgetown.edu/bitstream/handle/10822/515456/SACGHS_Final_Gene_Patents_Report_April2010.pdf?sequence=1.

24 Dan Vorhaus, "SACGHS Gene Patent Recommendations Still Controversial," *Genomics Law Report* (blog), February 8, 2010. Online at http://www.genomicslawreport.com/index.php/2010/02/08/sacghs-gene-patent-recommendations-still-controversial/; Donald Zuhn, "BIO Comes Out Swinging Against SACGHS Report—Updated," *Biotech & Pharma Patent Law & News* (blog), *PatentDocs*, February 4, 2010. Online at http://www.patentdocs.org/2010/02/bio-comes-out-swinging-against-sacghs-report.html.

25 Kathleen Sebelius, HHS Secretary, to Dr. Steven Teutsch, SACGHS, "Thank you for your work on *Gene Patents* . . ." letter, July 2, 2010. Online at http://osp.od.nih.gov/sites/default/files/Secretarys_letter_to_SACGHS_on_Patents_Report.pdf.

26 Vorhaus, "SACGHS Gene Patent Recommendations Still Controversial"; Donald Zuhn, "BIO Sends Letter on SACGHS Report to HHS Secretary Sebelius," *Biotech & Pharma Patent Law & News* (blog), *Patent Docs*, February 11, 2010. Online at http://www.patentdocs.org/2010/02/bio-sends-letter-on-sacghs-report-to-hhs-secretary-sebelius.html.

27 Joe Matal, "A Guide to the Legislative History of the America Invents Act: Part I of II," *Federal Circuit Bar Journal* 21, no. 3 (2012): 435–512.

28 Matal, "Guide to the Legislative History of the America Invents Act."

29 US Patent and Trademark Office, Public Hearing on Genetic Diagnostic Testing, Transcript, February 16, 2012. Online at http://www.uspto.gov/sites/default/files/aia_implementation/120216-genetic_transcript.pdf, 10.

30 Ibid., 70.

31 Ibid., 119.

32 Ibid., 165.

33 Ibid., 36–37.

34 US Patent and Trademark Office, "Public Comments and Notice of Public Hearings on Genetic Diagnostic Testing." Online at http://www.uspto.gov/patent/laws-and-regulations/america-invents-act-aia/public-comments-and-notice-public-hearings.

35 Food and Drug Administration, "Electronic Cigarettes and the Public Health: Public Workshop," Regulations.gov Web site, last modified March 20, 2015. Online at http://www.regulations.gov/#!documentDetail;D=FDA-2014-N-1936-0003.

36 US Patent and Trademark Office, Report on Confirmatory Genetic Diagnostic Test Activity, September 2015.

37 Tania Simoncelli, "AMP V. MYRIAD: PRELIMINARY REFLECTIONS," *GeneWatch* 26, no. 2–3 (2013): 7.

38 Interview with ACLU staff #1, April 20, 2012.

39 I participated in one of these meetings on November 6, 2007.

40 Interview with PubPat staff #1, April 19, 2012.

41 Samuel Walker, *In Defense of American Liberties: A History of the ACLU*, 2nd ed. (Carbondale: Southern Illinois University Press, 1999).

42 Interview with ACLU staff #2, April 20, 2012.

43 Interview with ACLU staff #1, April 20, 2012.

44 Maren Klawiter, *The Biopolitics of Breast Cancer: Changing Cultures of Disease and Activism* (Minneapolis: University of Minnesota Press, 2008); Ulrike Boehmer, *The Personal and the Political: Women's Activism in Response to the Breast Cancer and AIDS Epidemics* (Albany, NY: SUNY Press, 2000); Karen Stabiner, *To Dance with the Devil: The New War on Breast Cancer* (New York: Delacorte Press, 1997).

45 ACLU staff #1, e-mail message to author, April 20, 2012.

46 Grody, "Regulatory Issues in Molecular Diagnostics."

47 Association of Molecular Pathology et al. v. Myriad Genetics and PTO, "Complaint," United States District Court Southern District of New York, May 12, 2009, 6, on file with author.

48 BCAction editorial staff with Selene Kay of the ACLU, "BCAction Joins Coalition to Challenge BRCA Gene Patents," *The Source* 106, June 21, 2009. Online at http://bcaction.org /2009/06/21/bca-joins-coalition-to-challenge-brca-gene-patents/.

49 Interview with ACLU staff #1, April 20, 2012.

50 Brief for the National Women's Health Network et al., AMP et al., v. Myriad, 569 U.S. ___ (2013) (No. 12-398), *In Support of Plaintiffs' Opposition to Defendants' Motion to Dismiss and in Support of Plaintiffs' Motion for Summary Judgment* (August 28, 2009); Brief for the American Medical Association et al., 569 U.S. ___ (2013) (No. 12-398), *In Support of Plaintiffs' Opposition to Defendants' Motion to Dismiss and in Support of Plaintiffs' Motion for Summary Judgment* (August 27, 2009); Sir John Sulston, "BRCA—Statement of Support," American Civil Liberties Union, May 12, 2009. Online at https://www.aclu.org/free -speech/brca-statement-support-sir-john-sulston.

51 Wayne W. Grody, "Expert Declaration," Association of Molecular Pathology et al. v. US PTO and Myriad Genetics, Civil Action No. 09-4515. Online at https://www.aclu.org/files /pdfs/freespeech/brca_Grody_declaration_20090826.pdf; Genau Gerard, "Expert Declaration," Association of Molecular Pathology et al. v. US PTO and Myriad Genetics, Civil Action No. 09-4515. Online at https://www.aclu.org/files/pdfs/freespeech/brca_Girard _declaration_20090826.pdf.

52 Shobita Parthasarathy, "Expert Declaration," Association of Molecular Pathology et al. v. US PTO and Myriad Genetics, Civil Action No. 09-4515. Online at https://www.aclu.org/ files/pdfs/freespeech/brca_Parthasarathy_declaration_20090826.pdf.

53 Association of Molecular Pathology et al. v. Myriad Genetics and PTO, "Complaint," United States District Court Southern District of New York, May 12, 2009, 2.

54 Ibid., 29.

55 Association of Molecular Pathology et al. v. USPTO et al., *Opinion*, Case 1:09-cv-04515ORWS 09 Civ. 4515 (2010).

56 John Conley, "Pigs Fly: Federal Court Invalidates Myriad's Patent Claims," *Genomics Law Report* (blog), March 30, 2010. Online at http://www.genomicslawreport.com/index.php /2010/03/30/pigs-fly-federal-court-invalidates-myriads-patent-claims/; Jim Dwyer, "In Patent Fight, Nature, 1, Company, 0," *New York Times*, March 30, 2010; Association for Molecular Pathology et al. v. US PTO and Myriad Genetics, Inc., no. 09 Civ. 4515, 702 F. Supp. 2d 181 (S.D.N.Y. 2010).

57 Association of Molecular Pathology et al. v. US PTO and Myriad Genetics, No. 2010-1406, "Brief for the United States as Amicus Curiae in Support of Neither Party," Appeal from the United States District Court for the Southern District of New York, in case no. 09-CV-4515, Senior Judge Robert W. Sweet.

58 Duke Center for Genomic and Computational Biology (n.d.), "BRCA Resources," Audio file. Online at http://oralarguments.cafc.uscourts.gov/Audiomp3/2010-1406.mp3.

59 Matthew D. Henry and John L. Turner, "The Court of Appeals for the Federal Circuit's Impact on Patent Litigation," *Journal of Legal Studies* 35, no.1 (2006): 85–117.

60 United States Court of Appeals for the Federal Circuit Docket 2010-1406, decided July 29, 2011, https://www.aclu.org/files/assets/10-1406.pdf.

61 Ibid., 8.

62 Ibid., 28 of concurrent opinion.

63 Ibid.

64 Ibid.

65 Association of Molecular Pathology et al. v. Myriad Genetics, Supreme Court of the United States Hearing, Transcript, April 15, 2013. Online at http://www.supremecourt .gov/oral_arguments/argument_transcripts/12-398-amc7.pdf, 24.

66 Ibid., 11.

67 Association of Molecular Pathology et al. v. Myriad Genetics, 569 U.S. ___ (2013). Online at http://www.supremecourt.gov/opinions/12pdf/12-398_1b7d.pdf.

68 Brief for Eric S. Lander in Support of Neither Party as Amicus Curiae, Association of Molecular Pathology et al. v. Myriad Genetics, 569 U.S. ___ (2013) (No. 12-398).

69 Interview with Daniel Ravicher, executive director of the Public Patent Foundation, April 19, 2012.

70 OSGATA et al. v. Monsanto, "Complaint," No. 11-cv-2163-NRB (S.D.N.Y. June 1, 2011).

71 OSGATA et al. v. Monsanto Company and Monsanto Technology LLC, No. 11-cv-2163-NRB (S.D.N.Y. June 1, 2011), 55.

72 OSGATA et al. v. Monsanto, "Memorandum and Order," No 11-cv-2163-NRB (S.D.N.Y. February 24, 2012).

73 OSGATA et al. v. Monsanto, "Decision," No 11-cv-2163-NRB (S.D.N.Y. February 24 2012).

74 Ibid., 18.

75 Ibid.

76 Wil S. Hylton, "Who Owns This Body?" *Esquire*, January 29, 2007. Online at http:// www.esquire.com/news-politics/a1063/esq0601-june-genes-rev/; Dean Baker, "Current Drug-Patent System Is Bad Medicine" *Al Jazeera America*. Online at http://america .aljazeera.com/opinions/2014/11/drug-patents-pharmaceuticalindustrygenericsindia .html; Elisabeth Rosenthal, "The Soaring Cost of a Simple Breath," *New York Times*. Online at http://www.nytimes.com/2013/10/13/us/the-soaring-cost-of-a-simple-breath .html?pagewanted=all&_r=0; Tamar Haspel, "Unearthed: Are Patents the Problem?"

Washington Post, September 29, 2014. Online at http://www.washingtonpost.com/life-style/food/unearthed-are-patents-the-problem/2014/09/28/9bd5ca90-4440-11e4-9a15 -137aa0153527_story.html.

77 Ira Flatow, "Gene Patenting," *Science Friday*. Online at http://www.sciencefriday.com/ segment/12/11/2009/gene-patenting.html; CBS News, *Patented Genes* (April 4, 2010). Online at http://www.cbsnews.com/videos/patented-genes/; John Schwartz, "Cancer Patients Challenge the Patenting of a Gene," *New York Times* May 13, 2009.

78 Ginny Graves, "Who Owns Your Genes?" *Self*, October 16, 2010.

79 Dylan Mohan Gray, *Fire in the Blood*, directed by Dylan Mohan Gray (Mumbai: Spark-water India, 2012); *The Corporation*, directed by Mark Achbar and Jennifer Abbott (Vancouver, British Columbia, Canada: Big Picture Media Corp., 2005), DVD; *The World According to Monsanto*, directed by Marie-Monique Robin (Image et Compagnie, Arte France, and Office National du film du Canada, 2008); *Fast Food Nation*, directed by Richard Linklater (Recorded Picture Company, Participant Media, Fuzzy Bunny Films, and BBC films, 2006).

80 Robert Kenner, Elise Pearlstein, and Kim Roberts, *Food Inc.*, directed by Robert Kenner (Magnolia Pictures, Participant Media, and River Road Entertainment, 2009).

81 Eamon Murphy, "Bowman v. Monsanto: The Price We All Pay for Roundup Ready Seeds," *Daily Finance*, May 21, 2013. Online at http://www.dailyfinance.com/on/monsanto-gmo -roundup-ready-seeds-patents-food-prices/.

82 "Genes and Patents: More Harm Than Good?" *Economist*, April 15, 2010. Online at http:// www.economist.com/node/15905837.

83 Angelina Jolie, "My Medical Choice," *New York Times*, May 14, 2013. Online at http://www .nytimes.com/2013/05/14/opinion/my-medical-choice.html.

84 Neil Young, *The Monsanto Years* (New York: Reprise Records, 2015).

85 Samantak Ghosh, "Are All Genes Equal?" *Boston University Journal of Science and Technology Law* 20, no. 1 (2014): 1–23; "Recent Poll Results: How Divisive Gene Patenting Is," *GEN News Highlights*, March 12, 2012. Online at http://www.genengnews.com/gen-news -highlights/recent-poll-results-highlight-how-divisive-gene-patenting-is/81246482/.

86 Parthasarathy, *Building Genetic Medicine*.

87 Ibid.

88 British Society for Human Genetics, "Patenting of Human Gene Sequences and the EU Draft Directive," September 1997, Online at https://securehost11.zen.co.uk/bshg2/documents/official_docs/patent_eu.htm, on file with author; Steve Emmott, "No Patents on Life: The Incredible Ten-Year Campaign Against the European Patent Directive," in *Redesigning Life? The Worldwide Challenge to Genetic Engineering*, ed. Brian Tokar (New York: Zed Books, 2001).

89 Wendy Watson, "FAO Dr. Imogen Evans," e-mail message to author, June 1997, on file with author.

90 Clinical Molecular Genetics Society, "Gene Patents and Clinical Molecular Genetic Testing in the UK," 1999, on file with author.

91 *Directive 98/44/EC of the European Parliament and of the Council of 6 July 1998 on the Legal Protection of Biotechnological Inventions*, 1998 O.J. (L 213) 13–21.

92 Institut Curie, "Against Myriad Genetic's Monopoly on Tests for Predisposition to Breast and Ovarian Cancer, the Institut Curie Is Initiating an Opposition Procedure with the European Patent Office," September 12, 2001, on file with author.

93 William E. Bird, Bird & Goen, to European Patent Office, "Facts and Submissions," On behalf of the Danish center for clinical genetics et al., letter in patent prosecution of EP705903, 7, EPO Register Plus database, on file with author.

94 European Patent Office, "Minutes of the Oral Proceedings Before the Opposition Division," EP0705903, Application No. 953056058, January 24, 2005, 8; Brigitte Behrens on behalf of Greenpeace, "Einspruch gegen Patent EP 0705903 Patent auf Gen für Brustkrebs [Patent on the breast cancer gene] (BRCA1)," February 22, 2002, 1. EPO Register Plus database, on file with author.

95 Vincenzo Scuteri for the Materdomani Hospital to the European Patent Office, "Annex 1 Facts and Arguments: Grounds for Opposition," Patent no. EP699754, 6. EPO Register Plus database, on file with author.

96 Florianne Koechlin, Blauen-Institute, to the European Patent Office, "Einspruch gegen das Patent EP 0705902," on behalf of the Swiss Social Democrat Party, November 28, 2001, 7. EPO Register Plus database, on file with author.

97 European Patent Office, "Minutes of the Oral Proceedings Before the Opposition Division," Patent No. EP-B-0705902, Application 95 305 601.7, January 19, 2005. EPO Register Plus database, on file with author.

98 Oksana Mitnovestski and D. Nicol, "Are Patents for Methods of Medical Treatment Contrary to the Ordre Public and Morality or 'Generally Inconvenient'?" *Journal of Medical Ethics* 30, no. 5 (2004): 470–75.

99 Bird, Bird & Goen, to European Patent Office, "Facts and Submissions," 5.

100 Citizens' Enquiries Unit, e-mail message to Kathryn Reeves, author's research assistant, December 2, 2014.

101 European Parliament, "European Parliament Resolution on the Patenting of BRCA1 and BRCA2 ('Breast Cancer') Genes," Joint Motion for a Resolution, October 3, 2001. Online at http://www.europarl.europa.eu/sides/getDoc.do?type=MOTION&reference=P5-RC-2001 -0633&language=EN.

102 Ibid.

103 Ibid.

104 Meredith Wadman, "Testing Time for Gene Patent as Europe Rebels," *Nature* 413, no. 6854 (2001): 443; Laura Slattery, "Human Body Proves a Rich Vein for Gene Firms," *Irish Times*, November 16, 2001, 53; Christiane Wirtz, "Wem gehört das menschliche Gen? [Who owns the human genome?]" *Süddeutsche Zeitung*, July 23, 2001, 22; Christina Berndt, "Brustkrebs-Gentest freigegeben," *Süddeutsche Zeitung*, May 19, 2004, 12.

105 European Patent Office, "Minutes of the oral proceedings before the Opposition Division," EP0705902, Application 953056017, January 19 2005, 9.

106 Ibid., 1.

107 European Patent Office, 'Myriad/Breast Cancer' Patent Revoked After Public Hearing (press release, May 18, 2004), on file with author.

108 European Patent Office, "Interlocutory Decision in Opposition Proceedings (Article 106(3) EPC)," Patent No. 0705902, Application No. 95305605.8, June 9, 2005, 16.

109 European Patent Office, "Interlocutory Decision in Opposition Proceedings (Article 106(3) EPC)," Patent No. 0705902, Application No. 95305605.8, June 9, 2005, 50.

110 European Patent Office, "Statement by the European Patent Office concerning the Resolution of the European Parliament of 4 October 2001 on the patenting of BRCA1 and BRCA2 ('breast cancer') genes," October 17 2001. Online at http://www.europarl.europa

.eu/sides/getDoc.do?pubRef=-//EP//TEXT+MOTION+P5-RC-2001-0633+0+DOC+XML+V0//EN&language=mt.

111 Arthur Schaffer, "Method for Breeding Tomatoes Having Reduced Water Content and Product of the Method," EPO Patent EP1211926B1, filed July 4, 2000, and issued September 29, 2003; Richard Mithen and Kathy Faulkner, "Method for Selective Increase of the Anticarcinogenic Glucosinolates in Brassica Species," EPO Patent EP1069819B1, filed April 8, 1999, and issued July 24, 2002.

112 No Patents on Seeds, 2014, "European Patents on Plants and Animals—Is the Patent Industry Taking Control of Our Food?" Online at http://no-patents-on-seeds.org/sites/default/files/news/european_patents_on_plants_and_animals_2014_2.pdf.

113 "Stop Patents on Plants and Animals!" No Patents on Seeds Web site. Online at http://www.no-patents-on-seeds.org/.

114 "Opposition against the Monsanto Patents on Severed Broccoli," No Patents on Seeds Web site. Online at http://no-patents-on-seeds.org/en/recent-activities/opposition-against-monsanto-patent-severed-broccoli.

115 Christoph Then and Ruth Tippe, "The Future of Seeds and Food under the Growing Threat of Patents and Market Concentration," No Patents on Seeds Web site. Online at http://no-patents-on-seeds.org/sites/default/files/news/report_future_of_seed_en.pdf.

116 No Patents on Seeds, "Wem gehört der Brokkoli? [Who owns broccoli?]," Submitted in EP1069819 (2007 broccoli patent), EPO Register Plus database, on file with author.

117 Method for breeding tomatoes having reduced water content and product of the method, EPO case No. G0002/12 (decided March 25, 2015).

118 European Patent Office, "Scenarios for the Future," 2007 Report. Online at http://documents.epo.org/projects/babylon/eponet.nsf/0/63A726D28B589B5BC12572DB00597683/$File/EPO_scenarios_bookmarked.pdf, 2.

119 European Patent Office, "Scenarios for the Future."

120 Ibid., 121–22.

121 Ibid., 83.

122 Ibid., 78.

123 Ibid., 76.

124 Interview with EPO employee F, May 27, 2008.

125 Interview with EPO employee B, August 5, 2008.

126 Alison Brimelow, "Raising the Bar, Efficiency, Quality, and Stakeholder Incentives," Panel session at the 4th Annual Conference of the EPIP Association, September 25, 2009.

127 Ibid.

128 President Battistelli, European Patent Office, "Biotech Patents Subject to Exacting Standards" (press release, September 24, 2010), on file with author.

129 Steve Howe, "16 October 2012: EPO Clarity Objections: Raising the Bar Too High?" Reddie & Grose Web site. Online at http://www.reddie.co.uk/news-and-resources/ip-developments/epo-clarity-objections-raising-the-bar-too-high; Patent Law Firm von Bezold & Partner, "'Raising the Bar': Changes of European Patent Law," Statement, November 19, 2009. Online at http://pa-beier.de/raising_the_bar_part1.pdf.

130 United States Patent and Trademark Office, Department of Commerce, "Patent Quality Metrics for Fiscal Year 2017 and Request for Comments on Improving Patent Quality Measurement," *Federal Register* 81, no. 58 (2016): 16142–45.

131 Geertrui Van Overwalle, Esther van Zimmeren, Birgit Verbeure, and Gert Matthijs,

"Models for Facilitating Access to Patents on Genetic Inventions," *Nature Reviews Genetics* 7 (2006): 143–48.

132 Esther van Zimmeren and Geertrui Van Overwalle, "A Paper Tiger? Compulsory License Regimes for Public Health in Europe," *International Review of Intellectual Property and Competition Law (IIC)*, December 1, 2010. Online at http://ssrn.com/abstract=1717974.

133 Ibid.

134 "Public Discussion at European Parliament," No Patents on Seeds Web site. Online at https://no-patents-on-seeds.org/en/information/news/public-discussion-european -parliament.

135 "No patents on plants and animals!" No Patents on Seeds Web site. Online at http:// www.no-patents-on-seeds.org/en/archive/open-letter-members-european-parliament -and-european-commission.

136 Ibid.

137 European Parliament (2012). "European Parliament resolution of 10 May 2012 on the patenting of essential biological processes." (2012/2623(RSP)).

138 Ibid.

139 Press Release, European Patent Office, No European patents for essentially biological breeding processes (December 9, 2010), on file with author; Catherine Saez, "EPO Backs Patents on Conventional Plants: Broccoli, Tomato Cases Decided," *Intellectual Property Watch*, April 1, 2015.

140 Christoph Then and Ruth Tippe, "European Patent Office at Crossroads: Report— Patents on Plants and Animals Granted in 2011," No Patents on Seeds (March 2012). Online at http://no-patents-on-seeds.org/sites/default/files/news/npos_patente_report _march_2012_en.pdf.

141 Oliver Ladendorf, "Regulations on Essentially Biological Processes Amended," International Briefings 2013. Online at http://www.maiwald.eu/assets/Uploads/ladendorf2013 10.pdf.

142. "Netherlands: No Monopolization in the Dutch Plant Breeding Industry." Fresh Plaza Web site. Online at http://www.freshplaza.com/article/110063/

CONCLUSION

1 Margaret Foster Riley and Richard A. Merrill, "Regulating Reproductive Genetics: A Review of American Bioethics Commissions and Comparison to the British Human Fertilisation and Embryology Authority," *Columbia Science and Technology Law Review* 6 (2005): 1–61; Michael Mulkay, *The Embryo Research Debate: Science and the Politics of Reproduction* (New York: Cambridge University Press, 1997).

2 David H. Guston, "Innovation Policy: Not Just a Jumbo Shrimp," *Nature* 454, no. 7207 (2008): 940–42; David H. Guston and Daniel Sarewitz, "Real-Time Technology Assessment," *Technology in Society* 24, nos. 1–2 (2002): 93–109.

3 Heidi Ledford, "CRISPR, the Disruptor," *Nature* 522 (2015): 20–24.

4 Glenn A. Bowen, "Document Analysis as a Qualitative Research Method," *Qualitative Research Journal* 9, no. 2 (2009): 27–40; N. K. Denzin, *The Research Act: A Theoretical Introduction to Sociological Methods* (New York: Aldine, 1970); M. Q. Patton, *Qualitative Evaluation and Research Methods*, 2nd ed. (Newbury Park, CA: Sage, 1990); E. W. Eisner, *The Enlightened Eye: Qualitative Inquiry and the Enhancement of Educational Practice* (Toronto: Collier Macmillan Canada, 1991).

BIBLIOGRAPHY

Abbott, Alison. "Fresh Hope for German Stem-Cell Patent Case." *Nature* 462, no. 7271 (2009): 265.

Abbott, Andrew. *The System of Professions: An Essay on the Division of Expert Labor.* Chicago: University of Chicago Press, 1988.

Adams, Mark D., Jenny M. Kelley, Jeannine D. Gocayne, Mark Dubnick, Mihael H. Polumeropoulos, Hong Xiao, Carl R. Merril, et al. "Complementary DNA Sequencing: Expressed Sequence Tags and Human Genome Project." *Science* 252 (1991): 1651–56.

Adrian, Johns. *Piracy: The Intellectual Property Wars from Gutenberg to Gates.* Chicago: University of Chicago Press, 2010.

Akrich, Madeleine, João Nunes, Florence Paterson, and Vololona Rabeharisoa, eds. *The Dynamics of Patient Organizations in Europe.* Paris, France: Presses de Mines, 2008.

Alavi, Atossa M. "Stem Cell Compromise: A Wolf in Sheep's Clothing, Constitutional Implication of the Bush Plan." *Health Matrix* 13, no. 1 (2003): 181.

Allen, Arthur. "Who Owns My Disease?" *Mother Jones.* Online at http://www.motherjones .com/politics/2001/11/who-owns-my-disease.

American Society of Human Genetics. "Position Paper on Patenting of Expressed Sequence Tags." November, 1991. Online at http://www.ashg.org/pdf/policy/ASHG_PS _November1991.pdf.

Andejeski, Yvonne, Isabelle T. Bisceglio, Kay Dickersin, Jean E. Johnson, Sabina I. Robinson, Helene S. Smith, Frances M. Visco, and Irene M. Rich. "Quantitative Impact of Including Consumers in the Scientific Review of Breast Cancer Research Proposals." *Journal of Women's Health & Gender-Based Medicine* 11, no. 4 (2002): 379–88.

Andrews, Edmund L. "Patents: Applications for Animals Up Sharply." *New York Times*, April 22, 1989, 36.

———. "Religious Leaders Prepare to Fight Patents on Genes." *New York Times*, May 13, 1995.

———. "US Seeks Patent on Genetic Codes, Setting Off Furor." *New York Times*, October 21, 1991. Online at http://www.nytimes.com/1991/10/21/us/us-seeks-patent-on-genetic -codes-setting-off-furor.html.

Andrews, Lori. "The Gene Patent Dilemma: Balancing Commercial Incentives with Health Needs." *Houston Journal of Health Law and Policy* 2, no. 1 (2002): 65–106.

Aoki, Keith. "Authors, Inventors and Trademark Owners: Private Intellectual Property and the Public Domain." *Columbia-VLA Journal of Law and the Arts* 18, nos. 1 and 2 (1992): 191–267.

———. *Seed Wars: Cases and Materials on Intellectual Property and Plant Genetic Resources.* Durham, NC: Carolina Academic Press, 2008.

Aron, Nan. *Liberty and Justice for All: Public Interest Law in the 1980s and Beyond.* Boulder, CO: Westview Press, 1989.

Attaran, Amir, "How Do Patents and Economic Policies Affect Access to Essential Medicines in Developing Countries?" *Health Affairs* 23, no. 3 (2004): 155–66.

Bagley, Margo A. "A Global Controversy: The Role of Morality in Biotechnology Patent Law." University of Virginia Law School Public Law and Legal Theory Working Paper Series, Paper 57 (2007).

———. "Patent First, Ask Questions Later: Morality and Biotechnology in Patent Law." *William and Mary Law Review* 45, no. 2 (2003): 469–547.

Baker, Dean, "Current Drug-Patent System Is Bad Medicine." *Al Jazeera America*. Online at http://america.aljazeera.com/opinions/2014/11/drug-patents-pharmaceuticalindustry genericsindia.html.

Baldwin, Peter. *The Copyright Wars: Three Centuries of Trans-Atlantic Battle* (Princeton, NJ: Princeton University Press, 2014).

Bechtold, Stefan. "Physicians as User Innovators." In *Intellectual Property at the Edge: The Contested Contours of IP*, edited by Rochelle C. Dreyfuss and Jane C. Ginsburg, 343–58. New York: Cambridge University Press, 2014.

Beers, Diane L. *For the Prevention of Cruelty: The History and Legacy of Animal Rights Activism in the United States*. Athens, OH: Swallow Press, 2006.

Bereznak, Alyssa. "The U.S. Government Has a Secret System for Stalling Patents." Online at https://www.yahoo.com/tech/the-u-s-government-has-a-secret-system-for-1042496 88314.html.

Berman, Elizabeth Popp. *Creating the Market University: How Academic Science Became an Economic Engine*. Princeton, NJ: Princeton University Press, 2011.

Berndt, Christina. "Brustkrebs-Gentest freigegeben." *Süddeutsche Zeitung*, May 19, 2004, 12.

Bernier, Maître J. B. "Droit Public and Ordre Public." *Transactions of the Grotius Society* 15 (1929): 83–91.

Bessen, James E. "Patent Thickets: Strategic Patenting of Complex Technologies." Last modified March 2003. Online at http://ssrn.com/abstract=327760.

Biagioli, Mario. "Patent Specification and Political Representation: How Patents Became Rights." In *Making and Unmaking Intellectual Property: Creative Production in Legal and Cultural Perspective*, edited by Mario Biagioli, Peter Jaszi, and Martha Woodmansee, 25–40. Chicago: University of Chicago Press, 2011.

Blaxter, Mark. "Patenting of Genes." *Nature* 355 (1992):104.

Block, Fred, and Matthew R. Keller. *State of Innovation: The U.S. Government's Role in Technology Development*. Boulder, CO: Paradigm Publishers, 2011.

Blumenthal, David. "Growing Pains for New Academic/Industry Relationships." *Health Affairs* 13, no. 3 (1994): 176–93.

Bodeker, Gerard. "Traditional Medical Knowledge, Intellectual Property Rights, & Benefit Sharing." *Cardozo Journal of International and Comparative Law* 11 (2003–2004): 785–814.

Boehm, Klaus. *The British Patent System I: Administration*. New York: Cambridge University Press, 1967.

Boehmer, Ulrike. *The Personal and the Political: Women's Activism in Response to the Breast Cancer and AIDS Epidemics*. Albany, NY: SUNY Press, 2000.

Bohlen, Jim. *Making Waves: The Origin and Future of Greenpeace*. Montreal, Quebec: Black Rose Books, 2000.

Bok, Derek. *Universities in the Marketplace: The Commercialization of Higher Education*. Princeton, NJ: Princeton University Press, 2004.

Boldrin, Michele, and David K. Levine. *Against Intellectual Monopoly*. New York: Cambridge University Press, 2010.

Bomberg, Elizabeth. *Green Parties and Politics in the European Union.* New York: Routledge, 1998.

Borrás, Susana. "The Governance of the European Patent System: Effective and Legitimate?" *Economy and Society* 35, no. 4 (2007): 594–610.

Bowen, Glenn A. "Document Analysis as a Qualitative Research Method." *Qualitative Research Journal* 9, no. 2 (2009): 27–40.

Bower, Hilary. "Whose Genes Are They Anyway? Genes Part 1: The Patent Debate." *Independent*, November 8, 1997, 52.

Boyle, James. "A Theory of Law and Information: Copyright, Spleens, Blackmail, and Insider Trading." *California Law Review* 80, no. 6 (1992): 1413–540.

———. *The Public Domain: Enclosing the Commons of the Mind.* New Haven, CT: Yale University Press, 2010.

Bracmort, Kelsi, and Richard K. Lattanzio. "Geoengineering: Governance and Technology Policy." *Congressional Research Service*, November 26, 2013. Online at https://www.fas.org/sgp/crs/misc/R41371.pdf.

Bremmer, Charles. "Euro-MPS Clear Way for Genetic Patents." *Times*, May 13, 1998.

Brenner, Barbara A. "Our Genes Belong to Us. For Now." Online at http://womensfoundationofcalifornia.com/2010/04/08/our-genes-belong-to-us-for-now/.

Briggle, Adam. *A Rich Bioethics: Public Policy, Biotechnology, and the Kass Council.* South Bend, IN: University of Notre Dame Press, 2010.

Brown, Mark, and David Guston. "Science, Democracy, and the Right to Research." *Science and Engineering Ethics* 15, no. 3 (2009): 351–66.

Bud, Robert. *Penicillin: Triumph and Tragedy.* New York: Oxford University Press, 2007.

———. *The Uses of Life: A History of Biotechnology.* Cambridge: Cambridge University Press, 1994.

Busby, Helen, Tamara Hervey, and Alison Mohr. "Ethical EU Law? The Influence of the European Group on Ethics in Science and New Technologies." *European Law Review* 33 (2008): 803–42.

Butler, Katherine, and Charles Arthur. "Anger as Europe Votes to 'Sell Off' Genes." *Independent*, May 13, 1998, 1.

Callon, Michel, and Vololona Rabeharisoa. "Research 'in the Wild' and the Shaping of New Social Identities." *Technology in Society* 25, no. 2 (2003): 193–204.

Cantor, Charles R., and Cassandra L. Smith. *Genomics: The Science and Technology behind the Human Genome Project.* Hoboken, NJ: Wiley, 2004.

Carbone, Julia, E. Richard Gold, Bhaven Sampat, Subhashini Chandrasekharan, Lori Knowles, Misha Angrist, and Robert Cook-Deegan. "DNA Patents and Diagnostics: Not a Pretty Picture." *Nature Biotechnology* 28 (2010): 784–91.

Caron-Flinterman, J. Francisca, Jacqueline E. W. Broerse, and Joske F. G. Bunders. "The Experiential Knowledge of Patients: A New Resource for Biomedical Research?" *Social Science & Medicine* 60, no. 11 (2005): 2575–84.

Carson, Rachel. *Silent Spring.* New York: Houghton Mifflin, 1962.

Cassier, Maurice. "Patents and Public Health in France: Exclusion of Drug Pand Granting of Pharmaceutical Process Patents Between the Two World Wars." *History and Technology* 24, no. 2 (2008): 153–71.

Caulfield, Timothy, and E. Richard Gold. "Genetic Testing, Ethical Concerns, and the Role of Patent Law." *Clinical Genetics* 57, no. 5 (2001): 370–75.

Chakrabarty, A. M. "Patenting of Life-Forms: From a Concept to Reality." In *Who Owns Life?* edited by David Magnus and Arthur Caplan. Amherst, NY: Prometheus Books, 2002.

Chien, Colleen V. "Patent Amicus Briefs: What the Courts' Friends Can Teach Us about the Patent System." Santa Clara University School of Law Legal Studies Research Paper Series, Paper No. 10-05 (2010): 1–28. Online at http://ssrn.com/abstract=1608111.

Childress, James F. "Deliberations of the Human Fetal Tissue Transplantation Research Panel." In *Biomedical Politics*, edited by Kathi Hanna, 215–57. Washington DC: National Academies Press, 1991.

Cho, Mildred K., Samantha Illangasekare, Meredith A. Weaver, Debra G. B. Leonard, and Jon F. Merz. "Effects of Patents and Licenses on the Provision of Clinical Genetic Testing Services." *Journal of Molecular Diagnostics* 5, no. 1 (2003): 3–8.

Cohen, Wesley M., and Stephen A. Merrill, eds. *Patents in the Knowledge-Based Economy*. Washington, DC: National Academies Press, 2003.

Colaianni, Alessandra, Subhashini Chandrasekharan, and Robert Cook-Deegan. "Impact of Gene Patents and Licensing Practices on Access to Genetic Testing and Carrier Screening for Tay-Sachs and Canavan Disease." *Genetics in Medicine* 23, no. 1s (2010): S5–S14.

Collins, Nick. "EU Court Blow to Stem Cell Research." *Daily Telegraph*, October 19, 2011, 18.

Conley, John. "Pigs Fly: Federal Court Invalidates Myriad's Patent Claims." *Genomics Law Report* (blog), March 30, 2010. Online at http://www.genomicslawreport.com/index.php /2010/03/30/pigs-fly-federal-court-invalidates-myriads-patent-claims/.

Connor, Steve. "Medicine Thrown Into Crisis by Stem Cell Ruling; EU Ban on Stem Cell Patents 'Stifles Medical Research.'" *Independent*, October 19, 2011. Online at http://www .independent.co.uk/news/science/medicine-thrown-into-crisis-by-stem-cell-ruling -2372562.html.

Cook-Deegan, Robert. *The Gene Wars: Science, Politics, and the Human Genome*. New York: W. W. Norton, 1994.

———, Subhashini Chandrasekharan, and Misha Angrist. "The Dangers of Diagnostic Monopolies." *Nature* 458 (2009): 405–6.

Cook, Trevor. "A European Perspective as to the Extent to Which Experimental Use, and Certain Other, Defences to Patent Infringement, Apply to Differing Types of Research." Online at http://www.ipeg.com/_UPLOAD%20BLOG/Experimental%20Use%20for%20IPI %20Chapters%201%20to%209%20Final.pdf.

Correa, Carlos. *Integrating Public Health Concerns into Patent Legislation in Developing Countries*. Geneva, Switzerland: South Centre, 2000.

Crespi, R. Stephen. "The European Biotechnology Directive Is Dead." *Trends in Biotechnology* 13, no. 5 (1995): 162–64.

Crispin Littlehales. "Dan Ravicher." *Nature Biotechnology* 26, no. 4 (2008): 369.

Daemmrich, Arthur. *Pharmacopolitics: Drug Regulation in the United States and Germany*. Chapel Hill: University of North Carolina Press, 2004.

Dalpé, Robert, Louise Bouchard, Anne-Julie Houle, and Louis Bédard. "Watching the Race to Find the Breast Cancer Genes." *Science, Technology, and Human Values* 28, no. 2 (2003): 187–216.

Daly, Christopher B. "The People's Bicentennial Commission: Slouching Towards the Economic Revolution." *Harvard Crimson*, April 28, 1975. Online at http://www .thecrimson.com/article/1975/4/28/the-peoples-bicentennial-commission-pif-you/.

Davies, Kevin, and Michael White. *Breakthrough: The Race to Find the Breast Cancer Gene*. New York: Wiley, 1996.

de Laet, Marianne. "Patents, Travel, Space: Ethnographic Encounters with Objects in Transit." *Environment and Planning D: Society and Space* 18, no. 2 (2000): 149–68.

Deere, Carolyn. *The Implementation Game: The TRIPS Agreement and the Global Politics of Intellectual Property Reform in Developing Countries*. New York: Oxford University Press, 2009.

Dent, Chris. "'Generally Inconvenient': The 1624 Statute of Monopolies as Political Compromise." *Melbourne University Law review* 33, no. 2 (2009): 16–56.

Denzin, N. K. *The Research Act: A Theoretical Introduction to Sociological Methods*. New York: Aldine, 1970.

"DNA Law Likely." *Chemical Week*, 1977, 26.

Dobyns, Kenneth W. *The Patent Office Pony: A History of the Early Patent Office*. Fredericksburg, VA: Sergeant Kirkland's Museum and Historical Society, 1994.

Doerflinger, Richard M. "Congressional Impasse on Human Cloning." *National Catholic Bioethics Quarterly*, Spring 2004. Online at http://www.ncbcenter.org/page.aspx?pid=941.

Doh, Jonathan P., and Terrence R. Guay. "Corporate Social Responsibility, Public Policy, and NGO Activism in Europe and the United States: An Institutional-Stakeholder Perspective." *Journal of Management Studies* 43, no. 1 (2006): 47–73.

Dowie, Mark, "Gods and Monsters." *Mother Jones*, January/February, 2004, 48–53.

Drahos, Peter. *The Global Governance of Knowledge: Patent Offices and Their Clients*. New York: Cambridge University Press, 2010.

Duffy, John F. "The FCC and the Patent System: Progressive Ideals, Jacksonian Realism, and the Technology of Regulation." *University of Colorado Law Review* 71 (2000): 1071.

Dunlap, Riley E., and Angela G. Mertig. "The Evolution of the US Environmental Movement from 1970 to 1990: An Overview." In *American Environmentalism: The U.S. Environmental Movement, 1970–1990*, edited Riley E. Dunlap and Angela G. Mertig, 1–10. Philadelphia: Taylor & Francis, 1992.

Dutfield, Graham. *Intellectual Property Rights and the Life Science Industries: Past, Present, and Future*. Hackensack, NJ: World Scientific, 2009.

———. "Patent Systems as Regulatory Institutions." *Indian Economic Journal* 54, no. 1 (2006): 62–90.

Dwyer, Jim. "In Patent Fight, Nature, 1, Company, 0." *New York Times*, March 30, 2010.

Eisenberg, Rebecca S. "Patenting the Human Genome." *Emory Law Journal* 39 (1990): 721–45.

———. "Public Research and Private Development: Patents and Technology Transfer in Government-Sponsored Research." *Virginia Law Review* 82, no. 8 (1996): 1662–727.

———. "Patents and the Progress of Science: Exclusive Rights and Experimental Use." *University of Chicago Law Review* 56, no. 3 (1989): 1017–86.

———. "Proprietary Rights and the Norms of Science in Biotechnology Research." *Yale Law Journal* 97, no. 2 (1987): 177–231.

Eisner, E. W. *The Enlightened Eye: Qualitative Inquiry and the Enhancement of Educational Practice*. Toronto: Collier Macmillan Canada, 1991.

Emmott, Steve. "No Patents on Life: The Incredible Ten-Year Campaign Against the European Patent Directive." In *Redesigning Life? The Worldwide Challenge to Genetic Engineering*, edited by Brian Tokar, 373–84. New York: Zed Books, 2001.

Enerson, Benjamin D. "Protecting Society from Patently Offensive Inventions: The Risk of Reviving the Moral Utility Doctrine." *Cornell Law Review* 89 (2004): 685–720.

Epstein, Steven. *Impure Science: AIDS, Activism, and the Politics of Knowledge*. Berkeley: University of California Press, 1996.

Esping-Andersen, Gøsta. *Welfare States in Transition: National Adaptations in Global Economies*. Thousand Oaks, CA: Sage, 1996.

Evans, Erin. "Constitutional Inclusion of Animal Rights in Germany and Switzerland: How Did Animal Protection Become an Issue of National Importance?" *Society and Animals* 18, no. 3 (2010): 231–50.

Ezrahi, Yaron. *The Descent of Icarus: Science and the Transformation of Contemporary Democracy*. Harvard University Press, 1990.

Finney, J. M. T., and George Walker. "Abrogate the Patent on Salvarsan." *Journal of the American Medical Association* 68, no. 21 (1917): 1573.

Flatow, Ira. "Gene Patenting." *Science Friday*. Online at http://www.sciencefriday.com /segment/12/11/2009/gene-patenting.html; CBS News, *Patented Genes* (April 4, 2010). Online at http://www.cbsnews.com/videos/patented-genes/.

Forde, M. "The 'Ordre Public' Exception and Adjudicative Jurisdiction Conventions." *International and Comparative Law Quarterly* 29, nos. 2–3 (1980): 259–73.

Fourcade, Marion. *Economists and Societies: Discipline and Profession in the United States, Britain, and France, 1890s to 1990s*. Princeton, NJ: Princeton University Press, 2009.

Fowler, Cary. "The Plant Patent Act of 1930: A Sociological History of Its Creation." *Journal of the Patent and Trademark Office Society* 82 (2000): 621–44.

"Fresh Debate Over the Life-form Ruling." *Chemical Week*, August 6, 1980, 47.

Froschmaier, Franz. "Some Aspects of the Draft Convention Relating to a European Patent Law." *International and Comparative Law Quarterly* 12, no. 3 (1963): 886–97.

Fullilove, Courtney. "The Archive of Useful Knowledge." PhD diss., Columbia University, 2009. ProQuest Dissertations & Theses database.

Gabriel, Joe. *Medical Monopoly: Intellectual Property Rights and the Origins of the Modern Pharmaceutical Industry*. Chicago: University of Chicago Press, 2014.

Gearheart, John, "New Potential for Human Embryonic Stem Cells." *Science* 282, no. 5391 (1998): 1061–62.

Geuna, Aldo, and Lionel J. J. Nesta. "University Patenting and Its Effects on Academic Research: The Emerging European Evidence." *Research Policy* 35 (2006): 790–807.

Ghosh, Samantak. "Are All Genes Equal?" *Boston University Journal of Science and Technology Law* 20, no. 1 (2014): 1–23.

Gieryn, Thomas F. *Cultural Boundaries of Science: Credibility on the Line*. Chicago: University of Chicago Press, 1999.

Gold, E. Richard, and Alain Gallochat. "The European Biotech Directive: Past as Prologue." *European Law Journal* 7, no. 3 (2001): 331–66.

Golden, J. M. "WARF's Stem Cell Patents and Tensions between Public and Private Sector Approaches to Research." *Journal of Law, Medicine and Ethics* 38, no. 2 (2010): 314–31.

Gorman, Joseph Bruce. *Kefauver: A Political Biography*. Oxford University Press, 1971.

Gottweis, Herbert. *Governing Molecules: The Discursive Politics of Genetic Engineering in Europe and the United States*. Cambridge, MA: MIT Press, 1998.

———. "Stem Cell Policies in the United States and in Germany: Between Bioethics and Regulation." *Policy Studies Journal* 30, no. 4 (2002): 444–69.

————, and Barbara Prainsack. "Emotion in Political Discourse: Contrasting Approaches to Stem Cell Governance in the USA, UK, Israel, and Germany." *Regenerative Medicine* 1, no. 6 (2006): 823–29.

————, Brian Salter, and Catherine Waldby. *The Global Politics of Human Embryonic Stem Cell Science.* New York: Palgrave Macmillan, 2009.

Graham, Stuart J. H., Bronwyn H. Hall, Dietmar Harhoff, and David C. Mowery. "Post-Issue Patent 'Quality Control': A Comparative Study of US Patent Re-Examinations and European Patent Oppositions." National Bureau of Economic Research Working Paper Series, 8807 (February 2002). Online at http://www.nber.org/papers/w8807.

Graves, Ginny. "Who Owns Your Genes?" *Self*, October 16, 2010.

Gray, Dylan Mohan. "Fire in the Blood." Directed by Dylan Mohan Gray. Mumbai: Sparkwater, India, September 2012.

Green, Daniel. "Parliament Scuppers a New Patents Directive." *Financial Times*, March 3, 1995, 4.

Green, Lawrence W., and Shawna L. Mercer. "Can Public Health Researchers and Agencies Reconcile the Push from Funding Bodies and the Pull from Communities?" *American Journal of Public Health* 91, no. 12 (2001): 1926–29.

Greene, Jenna. "He's Not Just Monkeying Around." *Legal Times*, August 16, 1999, 16–20.

Greene, Jeremy, and Scott H. Podolsky. "Reform, Regulation, and Pharmaceuticals—The Kefauver-Harris Amendments at 50." *New England Journal of Medicine* 367 (2012): 1481–83.

Grisinger, Joanna. *The Unwieldy American State: Administrative Politics since the New Deal.* New York: Cambridge University Press, 2012.

Grody, Wayne W. "Regulatory Issues in Molecular Diagnostics." In *Molecular Diagnostics: Techniques and Applications for the Clinical Laboratory*, edited by Wayne W. Grody, Robert M. Nakamura, Frederick L. Kiechle, and Charles Strom, 121–24. Burlington, MA: Academic Press, 2009.

————. "Regulatory Issues in Molecular Diagnostics." In *Molecular Diagnostics: Techniques and Applications for the Clinical Laboratory*, edited by Wayne W. Grody, Robert M. Nakamura, Frederick L. Kiechle, and Charles Strom, . Burlington, MA: Academic Press, 2009.

Grubb, Philip W., and Peter R. Thomsen. *Patents for Chemicals, Pharmaceuticals and Biotechnology: Fundamentals of Global Law, Practice, and Strategy.* 5th ed. New York: Oxford University Press, 2010.

Guellec, Dominique, and Bruno van Pottelsberghe de la Potterie. *The Economics of the European Patent System: IP Policy for Innovation and Competition.* New York: Oxford University Press, 2007.

Gupta, R. K., and L. Balasubrahmanyam. "The Turmeric Effect." *World Patent Information* 20, nos. 3–4 (1998): 185–91.

Guston, David H. "Innovation Policy: Not Just a Jumbo Shrimp." *Nature* 454, no. 7207 (2008): 940–42.

————, and Daniel Sarewitz. "Real-Time Technology Assessment." *Technology in Society* 24, nos. 1–2 (2002): 93–109.

Haar, Paul S. "Revision of the Paris Convention: A Realignment of Private and Public Interests in the International Patent System." *Brooklyn International Law Journal* 8, no. 1 (1982): 77–108.

Hahn, Roger. *The Anatomy of a Scientific Institution: The Paris Academy of Sciences, 1666–1803*. Berkeley: University of California Press, 1971.

Halliday, Samantha. "A Comparative Approach to the Regulation of Human Embryonic Stem Cell Research in Europe." *Medical Law Review* 12 (Spring 2004): 40–69.

Hanson, Jaydee, and Eric Hoffman. "Synthetic Biology: The Next Wave of Patents on Life." *GeneWatch*. Online at http://www.councilforresponsiblegenetics.org/genewatch/Gene WatchPage.aspx?pageId=310.

Harhoff, Dietmar, and Markus Reitzig. "Determinants of Opposition against EPO Patent Grants: The Case of Biotechnology and Pharmaceuticals." *International Journal of Industrial Organization* 22, no. 4 (2004): 443–80.

Harris, Richard. *The Real Voice*. New York: Macmillan, 1964.

Haspel, Tamar. "Unearthed: Are Patents the Problem?" *Washington Post*, September 29, 2014. Online at http://www.washingtonpost.com/lifestyle/food/unearthed-are-patents -the-problem/2014/09/28/9bd5ca90-4440-11e4-9a15-137aa0153527_story.html.

Hawkes, Nigel. "Euro MPs Turn Down Life-Form Patent Law." *The Times*, March 2, 1995.

Hayden, Cori. *When Nature Goes Public: The Making and Unmaking of Bio-Prospecting in Mexico*. Princeton, NJ: Princeton University Press, 2003.

Healy, Kieran. *Last Best Gifts: Altruism and the Market for Human Blood and Organs*. Chicago: University of Chicago Press, 2006.

Heidenheimer, Arnold, Hugh Heclo, and Carolyn Teich Adams. *Comparative Public Policy: The Politics of Social Choice in Europe and America*. New York: St. Martin's Press, 1990.

Heller, Michael A., and Rebecca S. Eisenberg. "Can Patents Deter Innovation? The Anticommons in Biomedical Research." *Science* 280, no. 5364 (1998): 698–701.

Hemasree, Chaliki, Starlene Loader, Jeffrey Levenkron, Wende Logan-Young, W. Jackson Hall, and Peter Rowley. "Women's Receptivity to Testing for a Genetic Susceptibility to Breast Cancer." *American Journal of Public Health* 85 (1995): 1133–35.

Henry, Matthew D., and John L. Turner. "The Court of Appeals for the Federal Circuit's Impact on Patent Litigation." *Journal of Legal Studies* 35, no.1 (2006): 85–117.

Hilaire-Pérez, Liliane. "Invention and the State in 18th-Century France." *Technology and Culture* 32, no. 4 (1991): 911–31.

Hilgartner, Stephen. "Novel Constitutions? New Regimes of Openness in Synthetic Biology." *BioSocieties* 7, no. 2 (2012): 188–207.

———. *Science on Stage: Expert Advice as Public Drama*. Stanford, CA: Stanford University Press, 2000.

Ho, Cynthia M. "Splicing Morality and Patent Law: Issues Arising from Mixing Mice and Men." *Washington University Journal of Law & Policy* 2 (2000): 247–85.

Holland, Suzanne, Karen Lebacqz, and Laurie Zoloth. *The Human Embryonic Stem Cell Debate: Science, Ethics, and Public Policy*. Cambridge, MA: MIT Press, 2001.

Holman, Christopher M. "The Impact of Human Gene Patents on Innovation and Access: A Survey of Human Gene Patent Litigation." *UMKC Law Review* 76, no. 2 (2007–2008): 295–361.

Horn, Robert Van, and Mattias Klaes. "Chicago Neoliberalism versus Cowles Planning." *Journal of the History of the Behavioral Sciences* 47, no. 3 (2011): 302–21.

Howard, Ted. *The P.B.C.: A History*. Washington, DC: The People's Bicentennial Commission, 1976.

————, and Jeremy Rifkin. *Who Should Play God?: The Artificial Creation of Life and What It Means for the Future of the Human Race*. New York: Dell, 1977.

Hughes, Sally S. "Making Dollars Out of DNA: The First Major Patent in Biotechnology and the Commercialization of Molecular Biology, 1974–1980." *Isis* 92, no. 3 (2001): 541–75.

Husserl, Gerhart. "Public Policy and Ordre Public." *Virginia Law Review* 25, no. 1 (1938): 37–67.

Hutchison, Clyde A. "DNA Sequencing: Bench to Bedside and Beyond." *Nucleic Acids Research* 35 (2007): 6227–37.

Hylton, Wil S. "Who Owns This Body?" *Esquire*, January 29, 2007. Online at http://www.esquire.com/news-politics/a1063/esq0601-june-genes-rev/.

Iles, Kevin. "A Comparative Analysis of the Impact of Experimental Use Exemptions in Patent Law on Incentives to Innovate." *Northwestern Journal of Technology and Intellectual Property* 4, no. 1 (2005): 61–82.

Ingram, Mrill, and Helen Ingram. "Creating Credible Edibles: The Organic Agriculture Movement and the Emergence of US Federal Organic Standards." In *Routing the Opposition: Social Movements, Public Policy, and Democracy*. Minneapolis: University of Minnesota Press, 2005.

Jaffe, Adam B., and Josh Lerner. *Innovation and its Discontents: How Our Broken Patent System Is Endangering Innovation and Progress, and What to Do About It*. Princeton, NJ: Princeton University Press, 2011.

Jaggar, Karuna. "From the ED: Historic Gene Patent Case and Rally at Supreme Court." *The Source* 118. Online at http://www.bcaction.org/2013/04/05/from-the-ed-historic-gene-patent-case-and-rally-at-supreme-court/.

Janis, Mark D. "Rethinking Reexamination: Toward a Viable Administrative Revocation System for US Patent Law." *Harvard Journal of Law and Technology* 11, no. 1 (1997): 1–122.

Jasanoff, Sheila. *Designs on Nature: Science and Democracy in Europe and the United States*. Princeton, NJ: Princeton University Press, 2005.

————. *The Fifth Branch: Science Advisers as Policymakers*. Cambridge, MA: Harvard University Press, 1990.

————. *Risk Management and Political Culture*. Thousand Oaks, CA: Sage, 1986.

————. *States of Knowledge: The Co-production of Science and the Social Order*. New York: Routledge, 2004.

————. "Taking Life: Private Rights in Public Nature." In *Lively Capital: Biotechnologies, Ethics, and Governance in Global Markets*, edited by Kaushik Sunder Rajan, 155–83. Durham, NC: Duke University Press, 2012.

————, and Sang-Hyun Kim. "Containing the Atom: Sociotechnical Imaginaries and Nuclear Power in the United States and South Korea." *Minerva* 47, no. 2 (2009): 119–46.

Jasper, James M., and Dorothy Nelkin. *The Animal Rights Crusade: The Growth of a Moral Protest*. New York: Free Press, 1991.

————, and Jane D. Poulsen. "Recruiting Strangers and Friends: Moral Shocks and Social Networks in Animal Rights and Anti-nuclear Protests." *Social Problems* 42, no. 4 (1995): 493–512.

Jeanne Loring. "A Patent Challenge for Human Embryonic Stem Cell Research." *Nature Reports Stem Cells*, November 8, 2007. Online at http://www.nature.com/stemcells/2007/0711/071108/full/stemcells.2007.113.html.

Jolie, Angelina. "My Medical Choice." *New York Times*, May 14, 2013. Online at http://www
.nytimes.com/2013/05/14/opinion/my-medical-choice.html.

Jonsen, Albert R. *The Birth of Bioethics*. New York: Oxford University Press, 1998.

Joppke, Christian. *Mobilizing Against Nuclear Energy: A Comparison of Germany and the
United States*. Berkeley: University of California Press, 1993.

Kahn, Alfred E. "Fundamental Deficiencies of the American Patent Law." *American Economic
Review* 30, no. 3 (1940): 475–91.

Kahn, Jonathan. *Race in a Bottle: The Story of BiDil and Racialized Medicine in a Post-
Genomic Age*. New York: Columbia University Press, 2014.

Kamstra, Gerald, Mark Döring, Nick Scott-Ram, Andrew Sheard, and Henry Wixon. *Patents
on Biotechnological Inventions: The E. C. Directive*. London: Sweet & Maxwell, 2002.

Kapczynski, Amy. "The Access to Knowledge: Mobilization and the New Politics of
Intellectual Property." *Yale Law Journal* 117, no. 804 (2008). Online at http://dx.doi.org/10
.2139/ssrn.1323525.

Karanović, Jelena. "Contentious Europeanization: The Paradox of Becoming European
Through Anti-Patent Activism." *Ethnos* 75, no. 3 (2010): 252–74.

Kean, Hilda. *Animal Rights: Political and Social Change in Britain since 1800*. London:
Reaktion Books, 1998.

Keay, Laura A. "Morality's Move within U.S. Patent Law: From Moral Utility to Subject
Matter." *AIPLA Quarterly Journal* 40, no. 3 (2012): 409–39.

Kelly, Susan E. "Public Bioethics and Publics: Consensus, Boundaries, and Participation
in Biomedical Science Policy." *Science, Technology, and Human Values* 28, no. 3 (2003):
339–64.

Kenner, Robert, Elise Pearlstein, and Kim Roberts. *Food Inc*. Directed by Robert Kenner.
Magnolia Pictures, Participant Media, and River Road Entertainment, July, 2009.

Kevles, Daniel J. "Ananda Chakrabarty Wins a Patent: Biotechnology, Law, and Society."
Historical Studies in the Physical and Biological Sciences 25, no. 1 (1994): 111–35.

———. "Of Mice & Money: The Story of the World's First Animal Patent." *Daedalus* 131, no. 2
(2002): 78–88.

Khan, B. Zorina. "An Economic History of Patent Institutions." Online at 2014, http://eh.net
/encyclopedia/an-economic-history-of-patent-institutions/.

Kinchy, Abby. *Seeds, Science, and Struggle: The Global Politics of Transgenic Crops*.
Cambridge, MA: MIT Press, 2012.

Kitschelt, Herbert P. "Political Opportunity Structures and Political Protest: Anti-Nuclear
Movements in Four Democracies." *British Journal of Political Science* 16, no. 1 (1986):
57–85.

Klawiter, Maren. *The Biopolitics of Breast Cancer: Changing Cultures of Disease and Activism*.
Minneapolis: University of Minnesota Press, 2008.

Kleinman, Daniel Lee. *Impure Cultures: University Biology and the World of Commerce*.
Madison: University of Wisconsin Press, 2003.

———. *Politics on the Endless Frontier: Postwar Research Policy in the United States*. Durham,
NC: Duke University Press, 1995.

Knox, Andrea, "The Great Gene Grab: Firms Toss Researchers for a Loop." *Philadelphia
Inquirer*, February 13, 2000.

Kranakis, Eda. "Patents and Power: European Patent-System Integration in the Context of
Globalization." *Technology and Culture* 48 (2007): 689–728.

Labonte, Ronald. "Global Right to Health Campaign Launched." *British Medical Journal* 331, no. 7511 (2005): 252.

Ledford, Heidi. "CRISPR, the Disruptor." *Nature* 522 (2015): 20–24.

Lee, Peter. "Toward a Distributive Commons in Patent Law." *Wisconsin Law Review* 2009, no. 4 (2009): 917–1016.

Lee, Soo-Chin, Barbara Bernhardt, and Kathy Helzlsouer. "Utilization of BRCA1/2 Genetic Testing in the Clinical Setting." *Cancer* 94, no. 6 (2002): 1876–85.

Lenoir, Noëlle. "Europe Confronts the Embryonic Stem Cell Research Challenge." *Science* 287, no. 5457 (2000): 1425–27.

Letwin, William L. "The English Common Law Concerning Monopolies." *University of Chicago Law Review* 21, no. 3 (1954): 355–85.

Levidow, Les, Susan Carr, and David Wield. "Genetically Modified Crops in the European Union: Regulatory Conflicts as Precautionary Opportunities." *Journal of Risk Research* 3, no. 3 (2000): 189–208.

Lewin, Tamar. "Move to Patent Cancer Gene Is Called Obstacle to Research." *New York Times*, May 21, 1996.

Lyle, Mark Hamilton. *The Gentle Subversive: Rachel Carson, Silent Spring, and the Rise of the Environmental Movement* (New York: Oxford University Press, 2007).

Machin, Nathan. "Prospective Utility: A New Interpretation of the Utility Requirement of Section 101 of the Patent Act." *California Law Review* 87, no. 2 (1999): 4231–56.

Machlup, Fritz. *An Economic Review of the Patent System.* Washington, DC: US Government Printing Office, 1958.

———, and Edith T. Penrose. "The Patent Controversy in the Nineteenth Century." *Journal of Economic History* 10, no. 1, 1950.

Mackenzie, John. "A Closer Look." *ABC World News Tonight*, with Peter Jennings, April 2, 1998.

MacLeod, Christine. "The Paradoxes of Patenting: Invention and Its Diffusion in 18th- and 19th-Century Britain, France, and North America." *Technology and Culture* 23, no. 4 (1991): 885–910.

———. *Inventing the Industrial Revolution: The English Patent System, 1660–1800.* New York: Cambridge University Press, 1988.

Magnus, David, Arthur L. Caplan, and Glenn McGee, eds. *Who Owns Life?* New York: Prometheus Books, 2002.

Mahajan, Manjari. "The Right to Health as the Right to Treatment: Shifting Conceptions of Public Health." *Social Research: An International Quarterly* 79, no. 4 (2012): 819–36.

Matal, Joe. "A Guide to the Legislative History of the America Invents Act: Part I of II." *Federal Circuit Bar Journal* 21, no. 3 (2012): 435–512.

May, Christopher, and Susan K. Sell. *Intellectual Property Rights.* Boulder, CO: Lynne Rienner, 2006.

Mazon, George W. "Products of Nature: The New Criteria." *Catholic University Law Review* 20 (1971): 783–90.

McCormick, Sabrina, Julia Brody, Phil Brown, and Ruth Polk. "Public Involvement in Breast Cancer Research: An Analysis and Model for Future Research." *International Journal of Health Services* 34, no. 4 (2004): 625–46.

McCrudden, Christopher. "Human Dignity and Judicial Interpretation of Human Rights." *European Journal of International Law* 19, no. 4 (2008): 655–724.

McGarey, Barbara M. and Annette Levey. "Patents, Products, and Public Health: An Analysis of the CellPro March-in Petition." *Berkeley Technology Law Journal* 14, no. 3 (1999): 1095–116.

Merges, Robert P. "Property Rights Theory and the Commons: The Case of Scientific Research." *Social Philosophy and Policy* 13, no. 2 (1996): 145–67.

———, and John Fitzgerald Duffy. *Patent Law and Policy: Cases and Materials.* 6th ed. (LexisNexis, 2013).

———, Peter S. Menell, and Mark A. Lemley. *Intellectual Property in the New Technological Age.* 6th ed. Aspen Publishers, 2012.

Merz, Jon F. "Disease Gene Patents: Overcoming Unethical Constraints on Clinical Laboratory Medicine." *Clinical Chemistry* 45, no. 3 (1999): 324–30.

———, Antigone G. Kriss, Debra G. B. Leonard, and Mildred K. Cho. "Diagnostic Testing Fails the Test." *Nature* 415 (2002): 577–79.

Mgbeoji, Ikechi. "The Juridical Origins of the International Patent System: Towards a Historiography of the Role of Patents in Industrialization." *Journal of the History of International Law* 5, no. 2 (2003): 403–22.

Miegel, Frederik, and Tobias Olsson. "From Pirates to Politicians: The Story of the Swedish File Sharers Who Became a Political Party." In *Democracy, Journalism, and Technology: New Developments in an Enlarged Europe*, edited by Nico Carpentier, Pille Pruulmann-Vengerfeldt, Kaarle Nordenstreng, Maren Hartmann, Peeter Vihalemm, Bart Cammaerts, Hannu Nieminen, and Tobias Olsson. Tartu, Estonia: Tartu University Press. Online at http://iamcr.org/democracy-journalism-and-technology.

Mirowski, Philip. *Science-Mart: Privatizing American Science.* Cambridge, MA: Harvard University Press, 2011.

Mitchell, R. C. "From Conservation to Environmental Movement: The Development of the Modern Environmental Lobbies." In *Governance and Environmental Politics*, edited by M. J. Lacey. Washington, DC: Wilson Center Press, 1989.

Mitnovetski, Oksana, and Dianne Nicol. "Are Patents for Methods of Medical Treatment Contrary to the Ordre Public and Morality or 'Generally Inconvenient'?" *Journal of Medical Ethics* 30, no. 5 (2004): 470–75.

Moore, Kelly. *Disrupting Science: Social Movements, American Scientists, and the Politics of the Military, 1945–1975.* Princeton, NJ: Princeton University Press, 2013.

Mossoff, Adam. "Who Cares What Thomas Jefferson Thought About Patents: Reevaluating the Patent 'Privilege' in Historical Context." *Cornell Law Review* 92 (2007): 953.

Mowery, David C., Richard R. Nelson, Bhaven N. Sampat, and Arvids Ziedonis. "The Growth of Patenting and Licensing by US Universities: An Assessment of the Effects of the Bayh-Dole Act on 1980." *Research Policy* 20, no. 1 (2001): 99–119.

Moy, R. Carl. "The History of the Patent Harmonization Treaty: Economic Self-Interest as an Influence." *John Marshall Law Review* 26, no. 3 (1993): 457–95.

Mueller, Janice M. "No 'Dilettante Affair': Rethinking the Experimental Use Exception to Patent Infringement for Biomedical Research Tools." *Washington Law Review* 76, no. 1 (2001): 1–66.

Mulkay, Michael, *The Embryo Research Debate: Science and the Politics of Reproduction.* New York: Cambridge University Press, 1997.

Murphy, Eamon. "Bowman v. Monsanto: The Price We All Pay for Roundup-Ready Seeds."

Daily Finance, May 21, 2013. Online at http://www.dailyfinance.com/on/monsanto-gmo
-roundup-ready-seeds-patents-food-prices/.

Murray, Fiona. "The Oncomouse That Roared: Hybrid Exchange Strategies as a Source of Distinction at the Boundary of Overlapping Institutions." *American Journal of Sociology* 116, no. 2 (2010): 341–88.

———. "The Stem-Cell Market: Patents and the Pursuit of Scientific Progress." *New England Journal of Medicine* 356, no. 23 (2007): 2341–43.

———, and Scott Stern, "Do Formal Intellectual Property Rights Hinder the Free Flow of Scientific Knowledge? An Empirical Test of the Anti-Commons Hypothesis." *Journal of Economic Behavior & Organization* 63, no. 4 (2007): 648–87.

Murray, Kali. *A Politics of Patent Law: Crafting the Participatory Patent Bargain.* New York: Routledge, 2012.

Musk, Elon. "All Our Patent Are Belong to You." TESLA (blog), June 12 2014. Online at http://www.teslamotors.com/blog/all-our-patent-are-belong-you.

Naik, Paul S. "Biotechnology Through the Eyes of An Opponent: The Resistance of Activist Jeremy Rifkin." *Virginia Journal of Law and Technology* 5, no. 2 (2000): 5–15.

National Breast Cancer Coalition. "Gene Patenting: Yes or No?" *Call to Action!* 4, nos. 3 and 4 (1997): 11.

Nelkin, Dorothy. *Controversy: Politics of Technical Decisions.* Thousand Oaks, CA: Sage, 1992.

Newman, Stuart A. "My Attempt to Patent a Human-Animal Chimera." *L'Observatoire de la Génétique* 27 (April–May 2006). Online at https://www.nymc.edu/sanewman/PDFs/L%27Observatorie%20Genetique_chimera.pdf.

Noll, Roger G. "The Politics and Economics of Implementing State-Sponsored Embryonic Stem-Cell Research." SIEPR Discussion Paper No. 04-28, June 2005. Online at http://www.researchgate.net/profile/Roger_Noll/publication/241767339_The_Politics_and_Economics_of_Implementing_State-Sponsored_Embryonic_Stem-Cell_Research/links/0f31753348f5b2b8bf000000.pdf.

Ottinger, Gwen. *Refining Expertise: How Responsible Engineers Subvert Environmental Justice Challenges.* New York: NYU Press, 2013.

Palombi, Luigi. "The Patenting of Biological Materials in the United States: A State of Policy Confusion." *Perspectives on Science* 23, no. 1 (2015): 35–65.

Parry, Bronwyn. *Trading the Genome: Investigating the Commodification of Bio-Information.* New York: Columbia University Press, 2004.

Parthasarathy, Shobita. "Breaking the Expertise Barrier: Understanding Activist Challenges to Science and Technology Policy Domains." *Science & Public Policy* 37, no. 5 (2010): 355–67.

———. *Building Genetic Medicine: Breast Cancer, Technology, and the Comparative Politics of Health Care.* Cambridge, MA: MIT Press, 2007.

Patents and the Constitution: Transgenic Animals: Hearing Before the H. Subcomm. on Courts, Civil Liberties, and the Admin. of Justice of the Comm. on the Judiciary, 100th Cong. 56 (1987).

Patton, M. Q. *Qualitative Evaluation and Research Methods.* 2nd ed. Newbury Park, CA: Sage, 1990.

Pederson, Thoru. "NIH Patent Rights." *Nature* 358, no. 6288 (1992): 617.

Penrose, Edith T. *The Economics of the International Patent System.* Baltimore, MD: Johns Hopkins University Press, 1951.

Perkins, John H. *Geopolitics and the Green Revolution: Wheat, Genes, and the Cold War.* New York: Oxford University Press, 1997.

Piper, S. Tina. "A Common Law Prescription for a Medical Malaise." In *The Common Law of Intellectual Property: Essays in Honour of Professor David Vaver,* edited by C. Ng, L. Bently, and G. D'Agostino, 143–60. Oxford: Hart Publishing, 2010.

Pogge, Thomas, Matthew Rimmer, and Kim Rubenstein. *Incentives for Global Public Health: Patent Law and Access to Essential Medicines.* New York: Cambridge University Press, 2010.

Pollard, Sidney. *The Integration of the European Economy Since 1815.* New York: Routledge, 2006.

Post, Robert C. " 'Liberalizers' Versus 'Scientific Men' in the Antebellum Patent Office." *Technology and Culture* 17, no. 1 (1976): 24–54.

Prainsack, Barbara, and Robert Gmeiner. "Clean Soil and Common Ground: The Biopolitics of Human Embryonic Stem Cell Research in Austria." *Science as Culture* 17, no. 4 (2008): 377–95.

Quigg, Donald J. "Animals—Patentability." *Official Gazette* 1077 (April 21, 1987): 24.

Rabeharisoa, Vololona, Tiago Moreira, and Madeleine Akrich. "Evidence-Based Activism: Patients' Organisations, Users' and Activist's Groups in Knowledge Society." *BioSocieties* 9, no. 2 (2014): 111–28.

Rader, Karen. *Making Mice: Standardizing Animals for American Biomedical Research, 1900–1955.* Princeton, NJ: Princeton University Press, 2004.

Rai, Arti. "Regulating Scientific Research: Intellectual Property Rights and the Norms of Science." *Northwestern University Law Review* 94, no. 1 (1999): 77–152.

———, and James Boyle. "Synthetic Biology: Caught between Property Rights, the Public Domain, and the Commons." *PLOS Biology* 5, no. 3 (2007): 389–93.

Rasmussen, Nicholas. *Gene Jockeys: Life Science and the Rise of Biotech Enterprise.* Baltimore, MD: Johns Hopkins University Press, 2014.

Resnik, David B. "Privatized Biomedical Research, Public Fears, and the Hazards of Government Regulation: Lessons from Stem Cell Research." *Health Care Analysis* 7, no. 3 (1999): 273–87.

Rifkin, Jeremy. *Own Your Own Job: Economic Democracy for Working Americans.* New York: Bantam Books, 1977.

———. "Patenting Forms of Animal Life: Is Nature Just a Form of Private Property." *New York Times,* April 26, 1987, 2.

———, and Randy Barber. *The North Will Rise Again: Pensions, Politics, and Power in the 1980s.* Boston: Beacon Press, 1978.

Riley, Margaret Foster, and Richard A. Merrill. "Regulating Reproductive Genetics: A Review of American Bioethics Commissions and Comparison to the British Human Fertilisation and Embryology Authority." *Columbia Science and Technology Law Review* 6 (2004–2005).

Rimmer, Matthew. "Myriad Genetics: Patent Law and Genetic Testing." *European Intellectual Property Review* 25, no. 1 (2003): 20–33.

Risenfeld, Stefan A. "Licenses and United States Industrial and Artistic Property Law." *California Law Review* 47, no. 1 (1959): 51–63.

Ritter, Dominique. "Switzerland's Patent Law History." *Fordham Intellectual Property, Media, and Entertainment Law Journal* 14, no. 2 (2004): 463–96.

Roberts, Leslie. "Genome Patent Fight Erupts." *Science* 254, no. 5029 (1991): 184–86.

————. "NIH Gene Patents, Round Two." *Science* 255, no. 5047 (1992): 912–13.

Rosenberg, Stephanie T. "Asserting the Primacy of Health over Patent Rights: A Comparative Study of the Processes That Led to the Use of Compulsory Licensing in Thailand and Brazil." *Developing World Bioethics* 14, no. 2 (2014): 83–91.

Rosenthal, Elisabeth, "The Soaring Cost of a Simple Breath." *New York Times.* Online at http://www.nytimes.com/2013/10/13/us/the-soaring-cost-of-a-simple-breath.html ?pagewanted=all&_r=0.

Rothman, David J. *Strangers at the Bedside: A History of How Law and Bioethics Transformed Medical Decision Making.* New York: Basic Books, 1992.

Rowohlt, Axel. "Patent Protest Erupts Over Human Stem Cell Research." *Deutsche Welle.* Last modified December 28, 2006. Online at http://www.dw.de/dw/article/0,,2292175 _page_0,00.html.

Ruess, Peter. "Accepting Exceptions? A Comparative Approach to Experimental Use in U.S. and German Patent Law." *Marquette Intellectual Property Law Review* 10, no. 1 (2006): 82–110.

Ruzek, Sheryl Burt. *The Women's Health Movement: Feminist Alternatives to Medical Control.* New York: Praeger, 1978.

Saez, Catherine. "EPO Backs Patents on Conventional Plants: Broccoli, Tomato Cases Decided." *Intellectual Property Watch*, April 1, 2015.

Salter, Brian. "Bioethics, Politics, and the Moral Economy of Human Embryonic Stem Cell Science: The Case of the European Union's Sixth Framework Programme." *New Genetics and Society* 26, no. 3 (2007): 269–88.

Sarewitz, Daniel. "How Science Makes Environmental Controversies Worse." *Environmental Science and Policy* 7, no. 5 (2004): 385–403.

Sargent, Lyman Tower. *Contemporary Political Ideologies: A Comparative Analysis.* Belmont, CA: Wadsworth, 2009.

Scheingold, Stuart A., and Austin Sarat. *Something to Believe In: Politics, Professionalism, and Cause Lawyering.* Stanford, CA: Stanford University Press, 2004.

Schiermeier, Quirin. "Germany Challenges Human Stem Cell Patent Awarded 'by Mistake.'" *Nature* 404 (2000): 3–4.

Schissel, Anna, Jon F. Merz, and Mildred K. Cho. "Survey Confirms Fears about Licensing of Genetic Tests." *Nature* 402 (1999): 118.

Schmieder, Sandra. "Scope of Biotechnology Inventions in the United States and in Europe: Compulsory Licensing, Experimental Use and Arbitration: A Study of Patentability of DNA-Related Inventions with Special Emphasis on the Establishment of an Arbitration-Based Compulsory Licensing System." *Santa Clara High Technology Law Journal* 21, no. 1 (2004): 164–234.

Schneider, Keith, "Agency and Congress Face Clash over Patenting of Animals." *New York Times*, July 23, 1987.

————. "Senator Asks Halt to Animal Patents." *New York Times*, May 15, 1987.

Schurman, Rachel, and William A. Munro. *Fighting for the Future of Food: Activists versus Agribusiness in the Struggle over Biotechnology.* Minneapolis: University of Minnesota Press, 2010.

Schwartz, John. "Cancer Patients Challenge the Patenting of a Gene." *New York Times*, May 13, 2009.

Sease, Edmund J. "Common Sense, Nonsense and the Compulsory License." *Journal of the Patent Office Society* 55, no. 4 (1973): 233–54.

Sell, Susan, and Christopher May. "Moments in Law: Contestation and Settlement in the History of Intellectual Property." *Review of International Political Economy* 8, no. 3 (2001): 467–500.

Shade, Hans. *Patents at a Glance: A Survey of Substantive Law and Formalities in 46 Countries*. Munich: Carl Heymanns Verlag KG, 1971.

Shevelow, Kathryn. *For the Love of Animals: The Rise of the Animal Protection Movement*. New York: Holt, 2009.

Shiva, Vandana. *Biopiracy: The Plunder of Nature and Knowledge*. Brooklyn, NY: South End Press, 1999.

Shulman, Seth. "PB&J Patent Punch-Up." *MIT Technology Review*, May 1, 2001.

Simoncelli, Tania. "AMP V. MYRIAD: PRELIMINARY REFLECTIONS." *GeneWatch* 26, nos. 2–3 (2013): 7.

Simpson, Alan, Nicholas Hildyard, and Sarah Sexton. "No Patents on Life: A Briefing on the Proposed EU Directive on the Legal Protection of Biotechnological Inventions." *Corner House Briefing 01*. Online at http://www.thecornerhouse.org.uk/resource/no-patents-life.

Skocpol, Theda. *Social Policy in the United States*. Princeton, NJ: Princeton University Press, 1995.

Slattery, Laura. "Human Body Proves a Rich Vein for Gene Firms." *Irish Times*, November 16, 2001, 53.

Smith, Andrew R. "Monsters at the Patent Office: The Inconsistent Conclusions of Moral Utility and the Controversy of Human Cloning." *DePaul Law Review* 53, no. 1 (2003): 159.

Smith, Deborah. "Approving GM People Was a 'Patent Mistake.'" *Sydney Morning Herald*, February 23, 2000, 3.

Spencer, Richard. "The German Patent Office." *Journal of the Patent Office Society* 31, no. 2 (1949): 79–87.

Sperling, Stefan. *Reasons of Conscience: The Bioethics Debate in Germany*. Chicago: University of Chicago Press, 2013.

Spriggs, James F., and Paul J. Wahlbeck. "Amicus Curiae and the Role of Information at the Supreme Court." *Political Research Quarterly* 50, no. 2 (1997): 365–86.

Stabiner, Karen. *To Dance with the Devil: The New War on Breast Cancer*. New York: Delacorte Press, 1997.

Stafford, Ned. "Oliver Brüstle: Fighting for Stem Cell Research in Germany." *Lancet* 366 (2005): 1521.

Stammer, Larry B., and Robert Lee Hotz. "Religious Leaders Oppose Genetic Patents." *Philadelphia Inquirer*, May 18, 1995.

Starr, Paul. *The Social Transformation of American Medicine: The Rise of a Sovereign Profession and the Making of a Vast Industry*. New York: Basic Books, 1984.

Steen, Kathryn. *The American Synthetic Organic Chemicals Industry: War and Politics, 1910–1930*. Chapel Hill: University of North Carolina Press, 2014.

Steunenberg, Bernard. "Playing Different Games: The European Parliament and the Reform of Codecision." In *The European Parliament: Moving toward Democracy in the EU*, edited by Bernard Steunenberg and Jacques Thomassen, 163–184 . New York: Rowman & Littlefield, 2002.

———, and Jacques Thomassen, eds. *The European Parliament: Moving toward Democracy in the EU*. New York: Rowman & Littlefield, 2002.

Strathern, Marilyn. "Potential Property, Intellectual Rights and Property in Persons." *Social Anthropology* 4, no. 1 (1996): 17–32.

Swanson, Kara. *Banking on the Body: The Market in Blood, Milk, and Sperm in Modern America*. Cambridge, MA: Harvard University Press, 2014.

———. "The Emergence of the Professional Patent Practitioner." *Technology and Culture* 50, no. 3 (2009): 519–48.

Tazi, Andrea. *Biotechnological Inventions and Patentability of Life: The US and European Experience*. Northampton, MA: Edward Elgar, 2015.

Teichmann, David L. "Regulation of Recombinant DNA Research: A Comparative Study." *Loyola of Los Angeles International and Comparative Law Review* 6, no. 1 (1983): 1–35.

Terry, Sharon. "Why Banning Patents Would Hurt Patients." *GeneWatch* 23, no. 5 (2010): 24–25.

Thompson, Charis. *Good Science: The Ethical Choreography of Stem Cell Research*. Cambridge, MA: MIT Press, 2013.

Thompson, Dennis. "The Draft Convention for a European Patent." *International and Comparative Law Quarterly* 22, no. 1 (1973): 51–82.

Tobbell, Dominique. *Pills, Power, and Policy: The Struggle for Drug Reform in Cold War America and Its Consequences*. Berkeley: University of California Press, 2012.

Trubek, Louis G. "Public Interest Law: Facing the Problems of Maturity." *University of Arkansas—Little Rock Law Review* 33, no. 4 (2011): 417–33.

Turchetti, Simone. "Patenting the Atom: The Controversial Management of State Secrecy and Intellectual Property Rights in Atomic Research." In *Knowledge Management and Intellectual Property: Concepts, Actors, and Practices from the Past to the Present*, ed. Stathis Arapostathis and Graham Dutfield. Northampton, MA: Edward Elgar, 2013.

Usselman, Steven W. *Regulating Railroad Innovation: Business, Technology, and Politics in America, 1840–1920*. New York: Cambridge University Press, 2002.

van Beuzekom, Brigitte, and Anthony Arundel. "OECD Biotechnology Statistics 2009." Online at http://www.oecd.org/sti/42833898.pdf.

van Overwalle, Geertrui, Esther van Zimmeren, Birgit Verbeure, and Gert Matthijs. "Models for Facilitating Access to Patents on Genetic Inventions." *Nature Reviews Genetics* 7 (2006): 143–48.

van Zimmeren, Esther, and Geertrui Van Overwalle. "A Paper Tiger? Compulsory License Regimes for Public Health in Europe." *International Review of Intellectual Property and Competition Law (IIC)*, December 1, 2010. Online at http://ssrn.com/abstract=1717974.

Vogel, David. *The Politics of Precaution: Regulating Health, Safety, and Environmental Risks in Europe and the United States*. Princeton, NJ: Princeton University Press, 2012.

Wadlow, C. "Strasbourg, the Forgotten Patent Convention, and the Origins of the European Patents Jurisdiction." *International Review of Intellectual Property and Competition Law* 41, no. 2 (2010): 123–49.

Wadman, Meredith. "Testing Time for Gene Patent as Europe Rebels." *Nature* 413 (2001): 443.

Walker, Jack. *Mobilizing Interest Groups in America: Patrons, Professions, and Social Movements*. Ann Arbor: University of Michigan Press, 1991.

Walker, Samuel. *In Defense of American Liberties: A History of the ACLU*. 2nd ed. Carbondale: Southern Illinois University Press, 1999.

Walsh, John P., Ashish Arora, and Wesley Cohen. "Effects of Research Tool Patents and Licensing on Biomedical Innovation." In *Patents in the Knowledge-Based Economy*. Washington, DC: National Academies Press, 2003.

———. "Working Through the Patent Problem." *Science* 299 (2003): 1021.

Walterscheid, Edward C. "Inherent or Created Rights: Early Views on the Intellectual Property Clause." *Hamline Law Review* 19, no. 1 (1995): 81.

Watts, Susan. "Backlash Blocks 'Invention' of Animals." *Independent*, November 30, 1992, 4.

Weiler, Joseph. *The Constitution of Europe: 'Do the New Clothes Have an Emperor?' and Other Essays on European Integration*. New York: Cambridge University Press, 1999.

Weiss, Rick. "A Crucial Human Cell Is Isolated, Multiplied; Embryonic Building Block's Therapeutic Potential Stirs Debate." *Washington Post*, November 6, 1998.

———. "Patent Sought on Making of Part-Human Creatures; Scientist Seeks to Touch Off Ethics Debate." *Washington Post*, April 2, 1998.

Wertz, Dorothy. "Embryo and Stem Cell Research in the United States: History and Politics." *Gene Therapy* 9, no. 11 (2002): 674–78.

Whitman, James Q. "The Two Western Cultures of Privacy: Dignity versus Liberty." *Yale Law Journal*, 113, no. 6 (2004): 1151–1221.

Wilkie, Tom. "Genetic Patent on Cancer Mouse Faces Opposition." *Independent*, January 13, 1993, 7.

Williams, Heidi. "Intellectual Property Rights and Innovation: Evidence from the Human Genome." *Journal of Political Economy* 121, no. 1 (2013): 1–27.

Wilson, Duncan. *The Making of British Bioethics*. Manchester, UK: Manchester University Press, 2014.

Wincott, Daniel. "Federalism and the European Union: The Scope and Limits of the Treaty of Maastricht." *International Political Science Review* 17, no. 4 (1996): 403–15.

Wirtz, Christiane. "Wem gehört das menschliche Gen?" [Who owns the human genome?] *Süddeutsche Zeitung*, July 23, 2001, 22.

Wright, Susan. *Molecular Politics: Developing American and British Regulatory Policy for Genetic Engineering, 1972–1982*. Chicago: University of Chicago Press, 1994.

Wynne, Brian. "Misunderstood Misunderstandings: Social Identities and Public Uptake of Science." In *Misunderstanding Science? The Public Reconstruction of Science and Technology*. New York: Cambridge University Press, 2004.

Zhang, Sarah. "Sean Parker Just Gave $250M to the Darling of Cancer Research." *Wired*, April 13, 2016. Online at http://www.wired.com/2016/04/sean-parker-just-gave-250m-darling-cancer-research/.

Zwerdling, Daniel. "Humanimals." Weekend *All Things Considered*, National Public Radio, April 5, 1998.

INDEX

Page numbers in *italics* refer to illustrations.